CONTENTS

CHAPTER 4

CHAPTER 5

CHAPTER 6

CLEP* BIOLOGY

Laurie Ann Callihan, Ph.D.

Edited by Stephen Hart, M.A.T.

Research & Education Association
Visit our website at: www.rea.com/studycenter

Planet Friendly Publishing
✔ Made in the United States
✔ Printed on Recycled Paper
Text: 10%   Cover: 10%
Learn more: www.greenedition.org

At REA we're committed to producing books in an Earth-friendly manner and to helping our customers make greener choices.

Manufacturing books in the United States ensures compliance with strict environmental laws and eliminates the need for international freight shipping, a major contributor to global air pollution.

And printing on recycled paper helps minimize our consumption of trees, water and fossil fuels. This book was printed on paper made with **10% post-consumer waste**. According to the Environmental Paper Network's Paper Calculator, by using this innovative paper instead of conventional papers, we achieved the following environmental benefits:

Trees Saved: 5 • Air Emissions Eliminated: 863 pounds
Water Saved: 813 gallons • Solid Waste Eliminated: 245 pounds

Courier Corporation, the manufacturer of this book, owns the Green Edition Trademark. For more information on our environmental practices, please visit us online at **www.rea.com/green**

Research & Education Association
61 Ethel Road West
Piscataway, New Jersey 08854
E-mail: info@rea.com

CLEP Biology with Online Practice Exams

Printed in the United States of America

Library of Congress Control Number 2012951223

ISBN-13: 978-0-7386-1102-0
ISBN-10: 0-7386-1102-6

REA® is a registered trademark of Research & Education Association, Inc.

CHAPTER 7

CHAPTER 8

APPENDIX

ABOUT OUR AUTHOR

Laurie Ann Callihan, Ph.D., received her graduate degree in Science Education from Florida State University, Tallahassee, Florida. She also holds a B.S. in Biology from Christian Heritage College, El Cajon, California. Dr. Callihan is currently a Postdoctoral Associate for the Promoting Science among English Language Learners (P-SELL) Scale-Up project in the Steinhardt School of Culture, Education, and Human Development at New York University. Her research areas include delivery of sound content in the science classroom (natural sciences), diversity, equity, and teacher professional development in science education. She was a 2009 University Fellow of Florida State University for the School of Teacher Education. She has fifteen years of classroom science teaching experience with expertise in content delivery of natural sciences, curriculum development, and teaching in diverse classroom settings. She is the author of a variety of curricula, books, and academic materials in the K-20 science venue. Her experiences also include speaking on issues of education and the family at educational conferences in the United States, Canada, and Mexico.

ABOUT OUR EDITOR

Stephen Hart is a writer and editor specializing in science, medicine, and technology. After majoring in zoology, Mr. Hart went on to earn his M.A.T. in Biology at the University of Washington. He has worked as an editorial consultant for Prentice Hall, Saunders, and the American Society of Plant Physiologists. His freelance writing has included assignments for Time-Life Books, *ABCNews.com*, and the NASA Astrobiology website.

ABOUT RESEARCH & EDUCATION ASSOCIATION

Founded in 1959, Research & Education Association (REA) is dedicated to publishing the finest and most effective educational materials—including study guides and test preps—for students in middle school, high school, college, graduate school, and beyond.

Today, REA's wide-ranging catalog is a leading resource for teachers, students, and professionals. Visit *www.rea.com* to see a complete list of all our titles.

ACKNOWLEDGMENTS

We would like to thank Ernestine Struzziero for her technical edit; Pam Weston, Publisher, for setting the quality standards for production integrity and managing the publication to completion; John Paul Cording, Vice President, Technology, for coordinating the design and development of the REA Study Center; Larry B. Kling, Vice President, Editorial, for his supervision of revisions and overall direction; Diane Goldschmidt and Michael Reynolds, Managing Editors, for coordinating development of this edition; Transcend Creative Services for typesetting this edition; and Weymouth Design and Christine Saul, Senior Graphic Designer, for designing our cover.

CHAPTER 1

Passing the CLEP Biology Exam

PASSING THE CLEP BIOLOGY EXAM

Congratulations! You're joining the millions of people who have discovered the value and educational advantage offered by the College Board's College-Level Examination Program, or CLEP. This test prep covers everything you need to know about the CLEP Biology exam, and will help you earn the college credit you deserve while reducing your tuition costs.

GETTING STARTED

There are many different ways to prepare for a CLEP exam. What's best for you depends on how much time you have to study and how comfortable you are with the subject matter. To score your highest, you need a system that can be customized to fit you: your schedule, your learning style, and your current level of knowledge.

This book, and the online tools that come with it, allow you to create a personalized study plan through three simple steps: assessment of your knowledge, targeted review of exam content, and reinforcement in the areas where you need the most help.

Let's get started and see how this system works.

Test Yourself & Get Feedback	Score reports from your online diagnostic and practice tests give you a fast way to pinpoint what you already know and where you need to spend more time studying.
Review with the Book	Study the topics tested on the CLEP exam. Targeted review chapters cover everything you need to know.
Improve Your Score	Armed with your score reports, you can personalize your study plan. Review the parts of the book where you're weakest and study the answer explanations for the test questions you answered incorrectly.

THE REA STUDY CENTER

The best way to personalize your study plan and focus on your weaknesses is to get feedback on what you know and what you don't know. At the online REA Study Center, you can access two types of assessment: a diagnostic exam and full-length practice exams. Each of these tools provides true-to-format questions and delivers a detailed score report that follows the topics set by the College Board.

Diagnostic Exam

Before you begin your review with the book, take the online diagnostic exam. Use your score report to help evaluate your overall understanding of the subject, so you can focus your study on the topics where you need the most review.

Full-Length Practice Exams

These practice tests give you the most complete picture of your strengths and weaknesses. After you've finished reviewing with the book, test what you've learned by taking the first of the two online practice exams. Review your score report, then go back and study any topics you missed. Take the second practice test to ensure you have mastered the material and are ready for test day.

If you're studying and don't have Internet access, you can take the printed tests in the book. These are the same practice tests offered at the REA Study Center, but without the added benefits of timed testing conditions and diagnostic score reports. Because the actual exam is computer-based, we recommend you take at least one practice test online to simulate test-day conditions.

AN OVERVIEW OF THE EXAM

The CLEP Biology exam consists of approximately 115 multiple-choice questions, each with five possible answer choices, to be answered in 90 minutes.

The exam covers the material one would find in a college-level general biology course. The exam stresses the broad field of the biological sciences organized into three major areas:

33% Molecular and Cellular Biology

34% Organismal Biology

33% Population Biology

For a list of subtopics covered on the exam, visit the College Board website.

ALL ABOUT THE CLEP PROGRAM

What is the CLEP?

CLEP is the most widely accepted credit-by-examination program in North America. CLEP exams are available in 33 subjects and test the material commonly required in an introductory-level college course. Examinees can earn from three to twelve credits at more than 2,900 colleges and universities in the U.S. and Canada. For a complete list of the CLEP subject examinations offered, visit the College Board website: *www.collegeboard.org/clep*.

Who takes CLEP exams?

CLEP exams are typically taken by people who have acquired knowledge outside the classroom and who wish to bypass certain college courses and earn college credit. The CLEP program is designed to reward examinees for learning—no matter where or how that knowledge was acquired.

Although most CLEP examinees are adults returning to college, many graduating high school seniors, enrolled college students, military personnel, veterans, and international students take CLEP exams to earn college credit or to demonstrate their ability to perform at the college level. There are no prerequisites, such as age or educational status, for taking CLEP examinations. However, because policies on granting credits vary among colleges, you should contact the particular institution from which you wish to receive CLEP credit.

Who administers the exam?

CLEP exams are developed by the College Board, administered by Educational Testing Service (ETS), and involve the assistance of educators from throughout the United States. The test development process is designed and implemented to ensure that the content and difficulty level of the test are appropriate.

When and where is the exam given?

CLEP exams are administered year-round at more than 1,200 test centers in the United States and can be arranged for candidates abroad on request. To find the test center nearest you and to register for the exam, contact the CLEP Program:

CLEP Services
P.O. Box 6600
Princeton, NJ 08541-6600
Phone: (800) 257-9558 (8 A.M. to 6 P.M. ET)
Fax: (609) 771-7088
Website: *www.collegeboard.org*

OPTIONS FOR MILITARY PERSONNEL AND VETERANS

CLEP exams are available free of charge to eligible military personnel and eligible civilian employees. All the CLEP exams are available at test centers on college campuses and military bases. Contact your Educational Services Officer or Navy College Education Specialist for more information. Visit the DANTES or College Board websites for details about CLEP opportunities for military personnel.

Eligible U.S. veterans can claim reimbursement for CLEP exams and administration fees pursuant to provisions of the Veterans Benefits Improvement Act of 2004. For details on eligibility and submitting a claim for reimbursement, visit the U.S. Department of Veterans Affairs website at *www.gibill.va.gov*.

CLEP can be used in conjunction with the Post-9/11 GI Bill, which applies to veterans returning from the Iraq and Afghanistan theaters of operation. Because the GI Bill provides tuition for up to 36 months, earning college credits with CLEP exams expedites academic progress and degree completion within the funded timeframe.

SSD ACCOMMODATIONS FOR CANDIDATES WITH DISABILITIES

Many test candidates qualify for extra time to take the CLEP exams, but you must make these arrangements in advance. For information, contact:

College Board Services for Students with Disabilities
P.O. Box 6226
Princeton, NJ 08541-6226
Phone: (609) 771-7137 (Monday through Friday, 8 A.M. to 6 P.M. ET)
TTY: (609) 882-4118
Fax: (609) 771-7944
E-mail: ssd@info.collegeboard.org

6-WEEK STUDY PLAN

Although our study plan is designed to be used in the six weeks before your exam, it can be condensed to three weeks by combining each two-week period into one.

Be sure to set aside enough time—at least two hours each day—to study. The more time you spend studying, the more prepared and relaxed you will feel on the day of the exam.

Week	Activity
1	Take the online Diagnostic Exam in the REA Study Center. The score report will identify topics where you need the most review.
2–4	Study the review chapters. Use your diagnostic score report to focus your study.
5	Take Practice Test 1 at the REA Study Center. Review your score report and re-study any topics you missed.
6	Take Practice Test 2 at the REA Study Center to see how much your score has improved. If you still got a few questions wrong, go back to the review and study any topics you may have missed.

TEST-TAKING TIPS

Know the format of the test. CLEP computer-based tests are fixed-length tests. This makes them similar to the paper-and-pencil type of exam because you have the flexibility to go back and review your work in each section.

Learn the test structure, the time allotted for each section of the test, and the directions for each section. By learning this, you will know what is expected of you on test day, and you'll relieve your test anxiety.

Read all the questions—completely. Make sure you understand each question before looking for the right answer. Reread the question if it doesn't make sense.

Annotate the questions. Highlighting the key words in the questions will help you find the right answer choice.

Read all of the answers to a question. Just because you think you found the correct response right away, do not assume that it's the best answer. The last answer choice might be the correct answer.

Work quickly and steadily. You will have 90 minutes to answer 115 questions, so work quickly and steadily. Taking the timed practice tests online will help you learn how to budget your time.

Use the process of elimination. Stumped by a question? Don't make a random guess. Eliminate as many of the answer choices as possible. By eliminating just two answer choices, you give yourself a better chance of getting the item correct, since there will only be three choices left from which to make your guess. Remember, your score is based only on the number of questions you answer correctly.

Don't waste time! Don't spend too much time on any one question. Remember, your time is limited and pacing yourself is very important. Work on the easier questions first. Skip the difficult questions and go back to them if you have the time.

Look for clues to answers in other questions. If you skip a question you don't know the answer to, you might find a clue to the answer elsewhere on the test.

Acquaint yourself with the computer screen. Familiarize yourself with the CLEP computer screen beforehand by logging on to the College Board website. Waiting until test day to see what it looks like in the pretest tutorial risks injecting needless anxiety into your testing experience. Also, familiarizing yourself with the directions and format of the exam will save you valuable time on the day of the actual test.

Be sure that your answer registers before you go to the next item. Look at the screen to see that your mouse-click causes the pointer to darken the proper oval. If your answer doesn't register, you won't get credit for that question.

THE DAY OF THE EXAM

On test day, you should wake up early (after a good night's rest, of course) and have breakfast. Dress comfortably, so you are not distracted by being too hot or too cold while taking the test. (Note that "hoodies" are not allowed.) Arrive at the test center early. This will allow you to collect your thoughts and relax before the test, and it will also spare you the anxiety that comes with being late. As an added incentive, keep in mind that no one will be allowed into the test session after the test has begun.

Before you leave for the test center, make sure you have your admission form and another form of identification, which must contain a recent photograph, your name, and signature (i.e., driver's license, student identification card, or current alien registration card). You will not be admitted to the test center if you do not have proper identification.

You may wear a watch to the test center. However, you may not wear one that makes noise, because it may disturb the other test-takers. No cell phones, dictionaries, textbooks, notebooks, briefcases, or packages will be permitted, and drinking, smoking, and eating are prohibited.

Good luck on the CLEP Biology exam!

CHAPTER 2

The Chemistry of Biology

THE CHEMISTRY OF BIOLOGY

CHEMICAL BASIS OF LIFE

Matter—Atoms, Elements, Molecules

The study of matter is known as **chemistry**. In order to understand why cells behave the way they do, you must first understand some basic chemistry—the basic components of matter and how they interact. Matter is made up of basic substances called **elements,** which cannot be broken down into any other substance. The simplest unit of an element that retains the element's characteristics is known as an **atom**. All matter is made up of atoms. The properties of matter are a result of the structure of atoms and their interaction with each other.

Each atom of a given element has a **nucleus** containing a unique number of protons and about the same number of neutrons. The nucleus is surrounded by electrons, which have much less mass than protons and neutrons. Elements are listed by atomic number on the Periodic Table of the Elements. The **atomic number** is the number of protons found in the nucleus of an atom of that element.

Protons, neutrons, and electrons also differ in their **charge**. In an uncharged atom, the number of protons is equal to the number of electrons. Electrons have much less mass than protons and neutrons. Electrons have a charge of −1, while protons have a charge of +1. Neutrons have no charge. The number of protons in the nucleus of an atom carries a positive charge equal to this number; that is, if an atom's nucleus contains 4 protons, the charge is +4. Since positive and negative charges attract, the positive charges of the nucleus attract an equal number of negatively charged electrons. In an uncharged atom, the number of protons is equal to the number of electrons.

13

Electrons travel freely in a three-dimensional space that may be called an electron cloud, which is made up of energy levels or energy shells and orbitals. Current models of the atom follow the principles of quantum mechanics, which predict the probabilities of an electron being in a certain area at a certain time. Although the term "orbital" is used, electrons do not orbit the nucleus like a planet orbiting a sun. The orbital (or electron cloud, or electron shell) represents an area of probability in which an electron might be found at a particular location.

Fig. 2-1 The Periodic Table of the Elements. Elements are listed by atomic number.

Periodic Table of the Elements

	1	2	3	4	5	6	7	8	9	10	11	12	13	14	15	16	17	18
1	1 H 1.01																	2 He 4.00
2	3 Li 6.94	4 Be 9.01											5 B 10.81	6 C 12.01	7 N 14.01	8 O 16.00	9 F 19.00	10 Ne 20.18
3	11 Na 22.99	12 Mg 24.30											13 Al 26.98	14 Si 28.09	15 P 30.97	16 S 32.07	17 Cl 35.45	18 Ar 39.95
4	19 K 39.10	20 Ca 40.08	21 Sc 44.96	22 Ti 47.87	23 V 50.94	24 Cr 52.00	25 Mn 54.94	26 Fe 55.84	27 Co 58.93	28 Ni 58.69	29 Cu 63.55	30 Zn 65.39	31 Ga 69.72	32 Ge 72.61	33 As 74.92	34 Se 78.96	35 Br 79.90	36 Kr 83.80
5	37 Rb 85.47	38 Sr 87.62	39 Y 88.91	40 Zr 91.22	41 Nb 92.91	42 Mo 95.94	43 Tc (98)	44 Ru 101.07	45 Rh 102.91	46 Pd 106.42	47 Ag 107.87	48 Cd 112.41	49 In 114.82	50 Sn 118.71	51 Sb 121.76	52 Te 127.60	53 I 126.90	54 Xe 131.29
6	55 Cs 132.91	56 Ba 137.33		72 Hf 178.49	73 Ta 180.95	74 W 183.84	75 Re 186.21	76 Os 190.23	77 Ir 192.22	78 Pt 195.08	79 Au 196.97	80 Hg 200.59	81 Tl 204.38	82 Pb 207.20	83 Bi 208.98	84 Po (209)	85 At (210)	86 Rn (222)
7	87 Fr (223)	88 Ra (226)		104 Rf (261)	105 Db (262)	106 Sg (263)	107 Bh (264)	108 Hs (265)	109 Mt (268)	110 Uun (269)	111 Uuu (272)	112 Uub (277)						

57 La 138.91	58 Ce 140.12	59 Pr 140.91	60 Nd 144.24	61 Pm (145)	62 Sm 150.36	63 Eu 151.96	64 Gd 157.25	65 Tb 158.93	66 Dy 162.50	67 Ho 164.93	68 Er 167.26	69 Tm 168.93	70 Yb 173.04	71 Lu 174.97
89 Ac (227)	90 Th 232.04	91 Pa 231.04	92 U 238.03	93 Np (237)	94 Pu (244)	95 Am (243)	96 Cm (247)	97 Bk (247)	98 Cf (251)	99 Es (252)	100 Fm (257)	101 Md (258)	102 No (259)	103 Lr (262)

Each shell has a particular amount of energy related to it, and is therefore also referred to as an **energy level**. Energy levels' names are designated by a number-letter combination (i.e., 1s, 2s, 2p, etc.). The quantum number of the energy level closest to the nucleus is 1 and progresses as the levels get farther from the nucleus (2, 3, etc.). The letter designation indicates the shape of that particular energy level. The energy level closest to the nucleus has the least energy related to it; the farthest has the most.

Each energy level has a limited capacity for holding electrons, and each energy level requires a different number of electrons to fill it. Lower energy levels (closer to the nucleus) have less capacity for electrons than those farther from the nucleus. Since electrons are attracted to the nucleus, electrons fill the electron shells closest to the nucleus (lowest energy levels) first. Once a given level is full, electrons start filling the next level out. The outermost occupied energy level of an element is called the **valence level**. The number of electrons in the valence level determines the combinations that this atom will be likely to make with other atoms. Atoms are more stable when every electron is paired and are most stable when their valence level is full. The tendency for an atom toward stability means that elements having unpaired or partially filled valence levels will easily gain or lose electrons in order to obtain a more stable configuration.

Common Elements

Atomic Number	Symbol	Common Name
1	H	Hydrogen
2	He	Helium
6	C	Carbon
7	N	Nitrogen
8	O	Oxygen
11	Na	Sodium
12	Mg	Magnesium
14	Si	Silicon
15	P	Phosphorous
16	S	Sulfur
17	Cl	Chlorine
19	K	Potassium
20	Ca	Calcium
24	Cr	Chromium
26	Fe	Iron
29	Cu	Copper
30	Zn	Zinc
47	Ag	Silver
53	I	Iodine
79	Au	Gold
80	Hg	Mercury
82	Pb	Lead
86	Rn	Radon

Chemical Bonds

The valence properties of atoms determine how they will bond with other atoms. Different types of bonds exist, and they form between the atoms that make up a molecule, between charged ions, or between different molecules with partial charges.

A **covalent bond** is formed between atoms when they share electrons. For instance, hydrogen has only one electron, which is unpaired, leaving the 1s valence shell one electron short of being full (2). Oxygen has 6 electrons in its valence level; it needs 2 more electrons for its valence level to be full (8). It is therefore easy for 2 hydrogen atoms to share their electrons with the oxygen, filling the valence levels of each.

Fig. 2-2   Hydrogen Bonds. Dotted lines represent hydrogen bonds between water molecules.

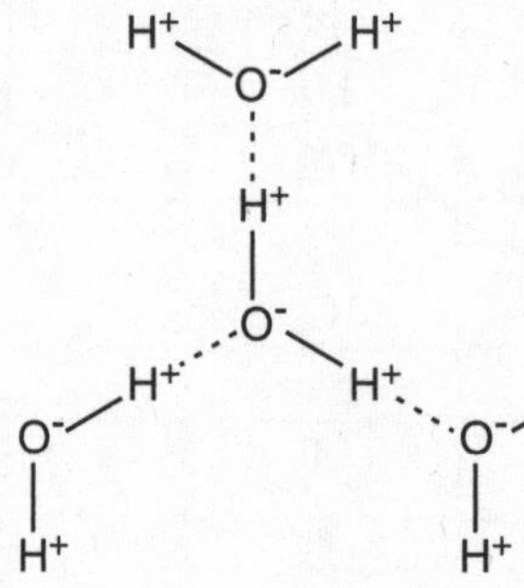

Covalent bonds are the strongest type of chemical bond. They result in a **molecule**, which is two or more atoms held together by covalently shared electrons. A **compound** is formed when two or more *different* atoms bond together chemically to form a unique substance (e.g., H_2O, CH_4). (In contrast, two atoms of the *same* element covalently bonded, e.g., O_2, results in a molecular element.)

Another type of bond involves charged atoms, which are called **ions**. An uncharged atom may be more stable if its valence level loses one or more electrons. It may lose one or more electrons to become a positively charged particle, or a positive ion. Similarly, an atom that gains one or more electrons to fill its valence level becomes a negative ion. When such an exchange occurs, the resulting oppositely-charged ions are attracted to each other and form an **ionic bond**. An example of a substance held together by ionic bonds is NaCl (sodium chloride or table salt). Ionic bonds are weaker than covalent bonds.

Some molecules have a weak, partial negative charge in one region of the molecule and a partial positive charge in another region. Molecules that have regions of partial charge are called **polar molecules**. For instance, water molecules (which have a net charge of 0) have a partial negative charge near the oxygen atom and a partial positive charge near each of the hydrogen atoms. Thus, when water molecules are close together, their positive regions are attracted to the negatively charged regions of nearby molecules; the negative regions are attracted to the positively charged regions of nearby molecules. The force of attraction between water molecules, shown above as a dotted line, is called a **hydrogen bond**. A hydrogen bond is a weak chemical bond that temporarily holds separate molecules together. For example, hydrogen bonds cause complementary strands of DNA to "zip" together to form a double strand.

Chemical Reactions

Chemical reactions occur when molecules interact with each other to form one or more molecules of another type. Chemical reactions that occur within cells provide energy, nutrients, and other products that allow the organism to function. Chemical reactions are symbolized by an equation where the reacting molecules (reactants) are shown on the left side and the newly formed molecules (products) are on the right side, with an arrow indicating the direction of the reaction. There are several categories of chemical reactions. Some chemical reactions are simple, such as the breakdown of a compound into its components (a decomposition reaction):

$$AB \rightarrow A + B$$

A simple combination reaction is the reverse of decomposition:

$$A + B \rightarrow AB$$

When one compound breaks apart and forms a new compound with a free reactant, it is called a replacement or displacement reaction:

$$AB + C \rightarrow AC + B$$

The Thermodynamics of Chemical Reactions

Chemical reactions may require an input of energy, or they may release energy. Reactions that require energy are called **endothermic** reactions.

Reactions that release energy are termed **exothermic**. Through endothermic reactions on the cellular level, living things are able to store chemical energy.

All chemical reactions are subject to the laws of thermodynamics. The first law of thermodynamics (also known as the law of conservation of matter and energy) states that matter and energy can neither be created nor destroyed. In other words, the sum of matter and energy of the reactants must equal that of the products. The second law of thermodynamics, or the law of increasing disorder (or entropy), asserts that all reactions increase energy disorder in a system, which tends to diminish its availability for cellular use. So, although we know from the first law that the energy must be equal on both sides of a reaction equation, reaction processes also tend to degrade the potential energy into a form that cannot perform any cellular work. The energy available to perform the work of a reaction is known as **free energy**.

Properties of Water

Water has unique characteristics that are important to the processes of life.

The transparent quality of water keeps it from disturbing processes within cells that require light (such as in photosynthetic and light-sensing cells).

The way water responds to temperature change is also unique. Most substances contract upon becoming a solid; however, water expands as it solidifies (freezes), forming a loose lattice structure (crystal). This crystalline form also makes frozen water (ice) less dense than liquid water. This accounts for lakes and other bodies of water freezing from the top first (the ice rises to the surface), insulating the water and organisms below from harsh temperature changes.

In addition, water has a high **specific heat**; it resists changes in temperature. The presence of water in an environment will tend to moderate otherwise harsh temperature changes. This is seen in the milder climates experiences in regions near large bodies of water.

Hydrogen bonds between water molecules also give water a high **surface tension**, allowing small particles, and even some organisms (such as the water strider), to rest on the surface.

Because water is a **polar** molecule, it is able to dissolve many types of organic and inorganic substances. This property promotes several biological processes such as muscle contraction, nerve stimulation, and transport across membranes (permeability).

Because water molecules are polar, certain types of chemicals dissociate in water. Some chemicals yield protons and others accept protons when dissolved in water. An **acid** is a chemical that donates protons (H^+ ions) when dissolved in water. A chemical that accepts protons (H^+ ions) or donates hydronium ions (OH^-) when dissolved in water is a **base**.

Because OH^- and H^+ ions combine to form water (H_2O), the presence of the base will lower the concentration of H^+ ions. Acids and bases tend to neutralize each other when dissolved together in water. The neutralization of an acidic solution with a basic solution produces water and a salt (an ionic compound) from the original acid and base molecules.

Acidity, then, is a measure of the concentration of H^+ ions in a solution and is described by the **pH scale**. The pH of a substance can range from 0 (the highest possible concentration) to 14 (the lowest possible). A pH of 7 is neutral (as is pure water). A substance with pH below 7 is acidic and one with a pH above 7 is basic (or alkaline). The pH of biological substances, such as blood and extracellular fluid, and of the water surrounding living things, is important to the chemistry of life processes.

Buffers

A **buffer** is an aqueous combination of a weak acid and its conjugate base, or a weak base and its conjugate acid. A buffer solution resists a change in pH when new H^+ or OH^- ions are added. The H^+ ions react with the weak base to form water; the OH^- ions react with the weak acid to form water. Maximum buffer capacity occurs when pH = pK_a for the weak acid. However, the buffering capacity will be exhausted if either the weak acid or the weak base is used up.

CHEMICAL STRUCTURE OF ORGANIC COMPOUNDS

Organic compounds are the building blocks of all living things. The special properties exhibited by the various types of organic molecules allow for the specialized functions within the cells and tissue of living things. **Organic compounds** are defined as those that contain carbon. Organic molecules may also include hydrogen, oxygen, nitrogen, sulfur, phosphorous, and some metal ions. Organic substances include many types of molecules active in biological processes (or biomolecules), such as carbohydrates, lipids, proteins, and nucleic acids. Many biological molecules, including DNA and protein, are large **polymers** made up of many of the same or similar subunits, called **monomers**.

Carbohydrates are up made of only carbon, hydrogen, and oxygen atoms, in varying ratios. The ratio of hydrogen to oxygen in carbohydrates is always 2:1, just as in water (H_2O)—thus the name *carbo* (carbon) *hydrate* (plus water). Sugars and starches are both forms of carbohydrates.

All carbohydrates are made up of a basic sugar unit called a **monosaccharide**, which usually contains 3 to 7 carbon atoms along with oxygen and hydrogen. The most common monosaccharides are hexoses (six-carbon sugars); they usually have a ring-shaped (or cyclic) structure. Glucose is one of the most important monosaccharides in human metabolism, as cells prefer this as an energy source.

Fig. 2-3   Structural formula of cyclic glucose (a common hexose).

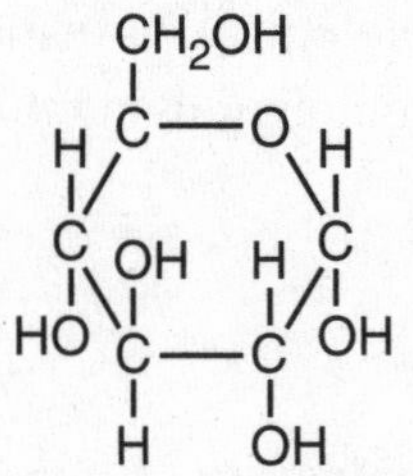

Two monosaccharide molecules may join together to form a **disaccharide** and liberate a molecule of water. Table sugar (sucrose) is a disaccharide of glucose and fructose (the most common monosaccharides). Glucose and fructose have the same chemical formula ($C_6H_{12}O_6$) but different arrangements of atoms; they are **isomers** of each other.

When three monosaccharides join together, the chain is then called a trisaccharide. When more than three monosaccharides chemically bond, the resultant molecule is known as a **polysaccharide**. Plant starches are the most familiar polysaccharides, and serve as energy storage molecules within the plant's cells, to be broken down when energy is needed. Plants also synthesize starches, which provide structure to their cells; the most common is a plant fiber known as **cellulose** (a long chain of water-insoluble polysaccharides).

Glycogen is a polysaccharide composed of many bonded glucose units. Many animals use glycogen as a short-term storage molecule for energy. In mammals, glycogen is found in muscle and liver tissue.

Lipids

Lipids are organic compounds composed of carbon, hydrogen, and oxygen. Unlike carbohydrates, the ratio of hydrogen to oxygen in lipids is always greater than 2: 1. Lipids include waxes, steroids, phospholipids, and fats.

Lipids are hydrophobic (water fearing) and will not dissolve in water. These various types of substances perform many functions within cells. Some form structural components of cell membranes (phospholipids), some provide moisture barriers (waxes), and others are primarily used to store energy (fats). Other lipids serve as vitamins or hormones.

Fig. 2-4 Production of Fat. The bonding of three fatty acids to one molecule of glycerol produces a fat (triglyceride) molecule and releases three water molecules. The release of water allows for the compacting of the high-energy fatty acids into a more concentrated fat molecule. This process is called dehydration synthesis.

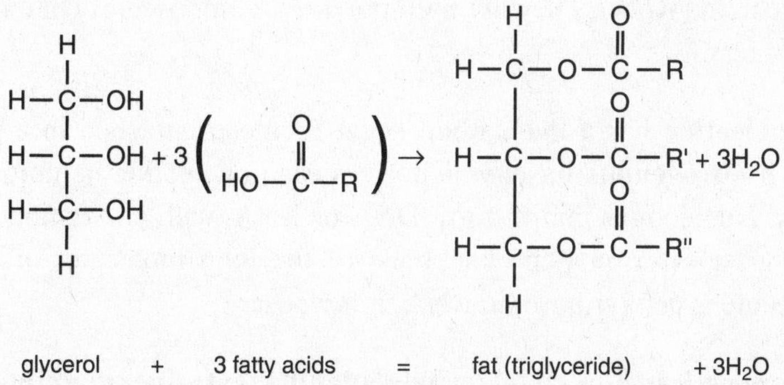

glycerol   +   3 fatty acids   =   fat (triglyceride)   + 3H$_2$O

Fats are highly efficient lipid molecules used for long-term energy storage. When an organism takes in more carbohydrates than are necessary for its current energy use, the excess is stored as fat molecules. These molecules "store" a lot of energy in a dense amount of mass, making it available to be used later. In addition to storing energy, fats also function in organisms to provide a protective layer that insulates internal organs and maintains heat within the body.

Proteins

Present in every living cell, **proteins** are large un-branched polymers made up of amino acid monomers. **Amino acids** are cyclical molecules that contain carbon, hydrogen, oxygen, nitrogen, and sometimes sulfur and phosphorous. In plants and animals, there are twenty common amino acids that can combine in various sequences to form thousands of different proteins. Amino acids are connected into chains by a water-releasing (dehydration) reaction that forms **peptide bonds**. For this reason, proteins may also be called **polypeptides**.

Proteins found in living things may have dozens or hundreds of amino acids. The long, linear strings of amino acids form unique shapes by folding up in various ways. The three-dimensional shape of a protein molecules is the characteristic that allows it to perform its specific functions within cells.

Enzymes are special proteins that act as catalysts for reactions. A catalyst is a substance that changes the speed of a reaction without being affected itself. Enzyme names have the suffix *-ase* (such as polymerase, lactase).

Nucleic Acids

There are two groups of nucleic acids: **deoxyribonucleic acid (DNA)** and **ribonucleic acid (RNA)**. They are both polymers composed of chains of nucleotide monomers.

Each **nucleotide** has a five-carbon sugar (pentose) attached to a phosphate group and a **nitrogenous base**, which gives each nucleotide its unique chemical identity. Nucleotides join to form DNA or RNA, with alternating sugar and phosphate groups forming the backbone of the long molecule. In DNA, the sugar molecule is deoxyribose; in RNA, it is ribose.

The **nitrogen bases** in DNA include **adenine**, **cytosine**, **guanine**, and **thymine** (A, C, G, and T). These parts of the nucleotides can hydrogen bond to form complementary pairs. That is, cytosine (C) and guanine (G) form hydrogen bonds and therefore pair together, while thymine (T) hydrogen bonds with adenine (A). Via hydrogen bonding, two complementary DNA strands pair up to form a shape like a twisted ladder. This double-helix structure of DNA was discovered and modeled by two scientists, James Watson and Francis Crick, in the 1950s. It is therefore known as the Watson-Crick model of DNA.

In RNA, thymine is replaced by the base **uracil** (U). RNA chains are generally single strands, but can pair up with a complementary DNA strand. **Complementary** strands have sequences of nucleotides that can **base-pair** (hydrogen bond) with each other.

DNA strand:	CTAATGTCATGTAT
Complementary DNA strand:	GATTACAGTACATA
Complementary RNA strand:	GAUUACAGUACAUA

Fig. 2-5   A DNA Molecule. Each molecule of DNA consists of a chain of nucleotides (each containing a phosphate group, a sugar, and a nitrogenous base). The chains bond together with hydrogen bonds forming a double-helix structure.

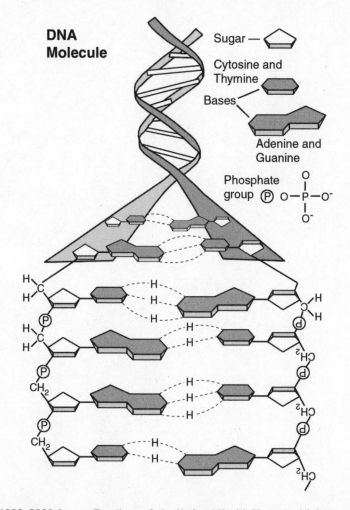

CHAPTER 3

Cellular and Molecular Biology

CELLULAR AND MOLECULAR BIOLOGY

CELL STRUCTURE AND FUNCTION

The **cell** is the smallest and most basic unit of structure for all living things (organisms). A single organism can be unicellular (consisting of just one cell) or multicellular (consisting of many cells). A multicellular organism may have many different types of cells that differ in structure to serve different functions. Individual cells may contain organelles that assist them with specialized functions. For example, muscle cells tend to contain more mitochondria (organelles that make energy available to the cells) since muscle requires the use of extra energy.

Scientists first began to describe cells after the invention of the light microscope in the mid-1600s. Antonie van Leeuwenhoek first observed tiny organisms (he called them "animalcules") with the use of microscopes. We now know these tiny organisms were one-celled bacteria. Robert Hooke was the first to use the term "cells" when he observed cell walls of dead cork under a light microscope.

In the mid-nineteenth century, two German scientists (Matthias Schleiden and Theodor Schwann) developed the **cell theory**. It consists of the following tenets:

1. All living things are made up of one or more cells.

2. Cells are the basic units of life.

3. All cells come from pre-existing cells.

These tenets of the cell theory developed by scientists over 150 years ago are still held today.

The light microscope is useful in examining most cells and some cell organelles (such as the nucleus). However, many cell organelles are very small and require the magnification and resolution power of an **electron microscope.**

There are two main types of cells: prokaryotic and eukaryotic. **Prokaryotes** are very simple; they have no nucleus or any other membrane-bound structures. The DNA in prokaryotic cells usually forms a single chromosome, which floats within the cytoplasm. Prokaryotic organisms are unicellular and include all bacteria.

Plant, fungi, and animal cells, as well as protozoa, are eukaryotic. **Eukaryotic** cells contain membrane-bound intracellular organelles (cell components that perform particular functions), including a nucleus. The DNA within eukaryotes is organized into chromosomes.

All cells are enclosed within the **cell membrane** (or plasma membrane). Near the center of each eukaryotic cell is the **nucleus**, which contains the chromosomes. Between the nucleus and the cell membrane, the cell is filled with **cytoplasm**. Since all of the organelles outside the nucleus but within the cell membrane exist within the cytoplasm, they are called **cytoplasmic organelles**.

The shape and size of cells can vary widely. The longest nerve cells (neurons) may extend over a meter in length with an approximate diameter of only 4–100 micrometers (1 millimeter = 1,000 micrometers, μm). A human egg cell may be 100 micrometers in diameter. The average size of a bacterium is 0.5 to 2.0 micrometers. However, most cells are between 0.5 and 100 micrometers in diameter. The size of a cell is limited by the ratio of its volume to its surface area. In the illustration below, note the variation of shape of cells within the human body:

Fig. 3-1 Varying Cell Types. The six sketches of human cell types show some of the diversity in shape and size among cells with varying functions. The sketches are not sized to scale.

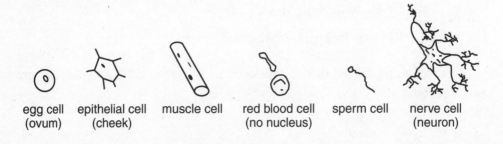

egg cell (ovum)   epithelial cell (cheek)   muscle cell   red blood cell (no nucleus)   sperm cell   nerve cell (neuron)

Animal cells differ in structure and function from cells of plants, fungi, and protists. For example, the photosynthetic cells have the added job of producing food, so they are equipped with specialized organelles called chloroplasts. Plant cells also have a central vacuole and cell walls, structures not found in animal cells.

Viruses are much smaller than even the smallest cells. Scientists do not agree as to whether viruses are actually alive. Although they can reproduce in host cells, they do not have the ability to conduct metabolic functions on their own. Virus structure consists of only a protein capsule, DNA, or RNA, and sometimes enzymes. Viruses survive and replicate by invading a living cell. The virus then utilizes the cell's mechanisms to reproduce itself, sometimes destroying the cell in the process.

Fig. 3-2   Viruses and Cells. Viruses are much smaller than cells, ranging from approximately 0.05–0.1 micrometers. The prokaryotic cell has no nucleus or other membrane-bound organelles and is approximately 1–10 micrometers in diameter. The eukaryotic cell has membrane-bound organelles, including a nucleus containing the chromosomes. Eukaryotic cells are approximately 10–100 micrometers in diameter.

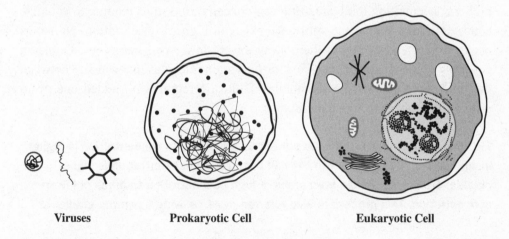

Viruses   Prokaryotic Cell   Eukaryotic Cell

PROPERTIES OF CELL MEMBRANES

The cell membrane is an especially important cell organelle with a unique structure that allows it to control movement of substances into and out of the cell. Made up of a fluid **phospholipid bilayer**, proteins, and carbohydrates, this extremely thin (approximately 80 angstroms) membrane can only be seen clearly with an electron microscope. The **selective permeability** of the cell membrane serves to manage the concentration of substances within the cell.

Fig. 3-3   Cell Membrane. A phospholipid bilayer with embedded globular proteins.

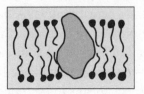

Substances can cross the cell membrane by passive transport, facilitated diffusion, or active transport. **Facilitated diffusion** does not require added energy, but it cannot occur without the help of specialized proteins. Transport requiring energy output from the cell is called **active transport**. During **passive transport,** substances freely pass across the membrane without the cell expending any energy.

Simple diffusion is one type of passive transport. **Diffusion** is the process whereby molecules and ions flow through the cell membrane from an area of higher concentration to an area of lower concentration (thus tending to equalize concentrations). Where the substance exists in higher concentration, collisions occur, which tend to propel them away toward lower concentrations. Diffusion generally is the means of transport for ions and molecules that can slip between the lipid molecules of the membrane. Diffusion requires no added energy to propel substances through a membrane.

Fig. 3-4   Diffusion. CO_2 diffuses out of the cell since its concentration is higher inside the cell. O_2 diffuses into the cell because its concentration is higher outside. Molecules diffuse from areas of high concentration to areas of lower concentration. This process is also referred to as "moving down the gradient."

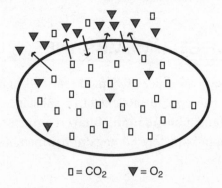

□ = CO_2 ▼ = O_2

Another type of passive transport is **osmosis,** a special process of diffusion occurring only with water molecules. Osmosis does not require the addition of any energy, but occurs when the water concentration inside the cell differs from the concentration outside the cell. The water on the side of the membrane with the highest water concentration will move through the membrane until the concentration is equalized on both sides. When the water concentration is equal inside and outside the cell, it is called isosmotic or isotonic. For instance, a cell placed in a salty solution will tend to lose water until the solution outside the cell has the same concentration of water molecules as the cytoplasm (the solution inside the cell).

Fig. 3-5   Osmosis. Water crosses the membrane into a cell that has a higher concentration of sugar molecules than the surrounding solution. Water crosses the membrane to leave the cell when there is a higher concentration of Na+ ions outside the cell than inside the cell.

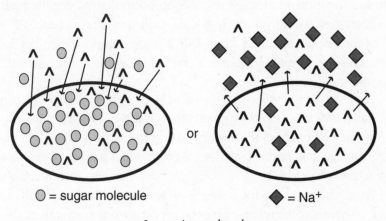

○ = sugar molecule   ◆ = Na+

∧ = water molecule

Facilitated diffusion allows for the transfer of substances across the cell membrane with the help of specialized proteins. These proteins, which are embedded in the cell membrane, provide channels for specific molecules or ions and transport them through the membrane. The special protein molecules allow the diffusion of molecules and ions that cannot otherwise pass through the lipid bilayer.

Fig. 3-6 Facilitated Diffusion. Specialized proteins embedded in the cell membrane permit passage of substances of a particular shape and size.

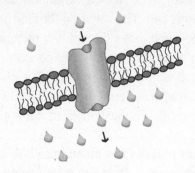

Active transport, like facilitated diffusion, requires membrane-bound proteins. Unlike facilitated diffusion, active transport uses energy to move molecules across a cell membrane against a concentration gradient (in the opposite direction than they would go under normal diffusion circumstances). With the addition of the energy obtained from ATP, a protein molecule embedded in the membrane changes shape and moves a molecule across the membrane against the concentration gradient.

Large molecules are not able to pass through the cell membrane, but may be engulfed by the cell membrane. **Endocytosis** is the process whereby large molecules (i.e., some sugars or proteins) are taken up into a pocket of membrane. The pocket pinches off, delivering the molecules, still inside a membrane sack, into the cytoplasm. This process, for instance, is used by white blood cells to engulf bacteria. **Exocytosis** is the reverse process, exporting substances from the cell.

CELL ORGANELLES OF PLANTS AND ANIMALS

Animal Cells

1. The **cell membrane (1a)** encloses the cell and separates it from the environment. It may also be called a plasma membrane. This membrane is composed of a double layer (bilayer) of phospholipids with globular proteins embedded within the layers. The membrane is extremely thin (about 80 angstroms; 10 million angstroms = 1 millimeter) and elastic. The combination of the lipid bilayer and the proteins embedded within it allows the cell to determine what molecules and ions can pass, and regulates the rate at which they enter and leave.

Endocytic vesicles (1b) form when the plasma membrane of a cell surrounds a particle outside the cell, then pinches off and releases into the cytoplasm a membrane-bound sac containing the particle. This process allows the cell to absorb larger molecules that otherwise would be unable to pass through the cell membrane, or that need to remain packaged within the cell.

Fig. 3-7   A Generalized Animal Cell (cross-section). Since there are many types of animal cells, the diagram is generalized. In other words, some animal cells will have all of these organelles, others will not. However, this illustration will give you a composite picture of the organelles within the typical animal cell. It is also important to note the function of each organelle. Each labeled component is explained by the corresponding text in the section below.

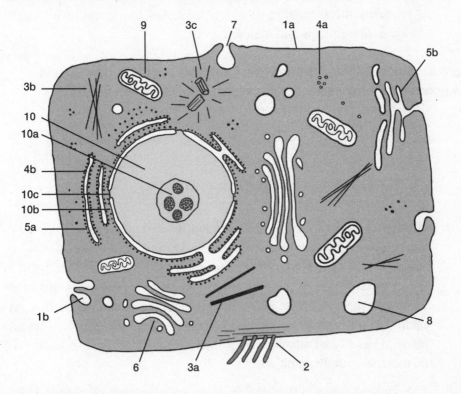

2. **Microvilli** are projections of the cell extending from the cell membrane. Microvilli are found in certain types of cells, for example, those involved in absorption (such as the cells lining the intestine). These filaments increase the surface area of the cell membrane, increasing the area available to absorb nutrients. They also contain enzymes involved in digesting certain types of nutrients.

3. The **cytoskeleton** provides structural support to a cell. **Microtubules (3a)** are long, hollow, cylindrical protein tubules, which give structure to the cell. These tubules are scattered around the edges of a cell and form a sort of loose skeleton or framework for the cytoplasm. Microtubules also are found at the base of cilia or flagella (organelles that allow some cells to move on their own) and give these organelles the ability to move. **Microfilaments (3b)** are double-stranded chains of proteins that serve to give structure to the cell. Together with the larger microtubules, microfilaments form the cytoskeleton, providing stability and structure. **Centrioles (3c)** are structural components of many cells and are particularly common in animal cells. Centrioles are tubes constructed of a geometrical arrangement of microtubules in a pinwheel shape. Their function includes the formation of new microtubules, but is primarily the formation of a structural skeleton around which cells divide during mitosis and meiosis.

Fig. 3.8   Cross-section of a Centriole. Centrioles are tubes constructed of a geometrical arrangement of microtubules in a pinwheel shape.

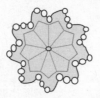

4. **Ribosomes** are the site of protein synthesis within cells. Ribosomes are composed of certain protein molecules and RNA molecules (ribosomal RNAs, or rRNAs). **Free ribosomes (4a)** float unattached within the cytoplasm. The proteins synthesized by free ribosomes are made for use in the cytoplasm, not within membrane-bound organelles. **Attached ribosomes (4b)** are attached to the ER (see No. 5). Proteins made at the site of attached ribosomes are destined for use within the membrane-bound organelles.

5. The **endoplasmic reticulum**, a large organization of folded membranes, is responsible for the delivery of lipids and proteins to certain areas within the cytoplasm (a sort of intra-cellular highway). **Rough endoplasmic reticulum or RER (5a)** has attached ribosomes. In addition to packaging and transporting materials within the cell, the RER is instrumental to protein synthesis. **Smooth endoplasmic reticulum or SER (5b)** is a network of membranous channels. Smooth endoplasmic

reticulum does not have attached ribosomes. The endoplasmic reticulum is responsible for processing lipids, fats, and steroids, which are then packaged and dispersed by the Golgi apparatus.

6. The **Golgi apparatus** (also known as Golgi bodies or the Golgi complex) is instrumental in the storing, packaging, and shipping of proteins. The Golgi apparatus looks much like stacks of hollow pancakes and is constructed of folded membranes. Within these membranes, cellular products are stored, or packaged by closing off a bubble of membrane with the proteins or lipids inside. These packages are shipped (via the endoplasmic reticulum) to the part of the cell where they will be used, or to the cell membrane for secretion from the cell.

7. **Secretory vesicles** are packets of material packaged by either the Golgi apparatus or the endoplasmic reticulum. Secretory vesicles carry substances produced within the cell (a protein, for example) to the cell membrane. The vesicle membrane fuses with the cell membrane in a process called **exocytosis**, allowing the substance to leave the cell.

8. **Lysosomes** are membrane-bound organelles containing digestive enzymes. Lysosomes break down unused material within the cell, damaged organelles, or materials absorbed by the cell for use.

9. **Mitochondria** are centers of cellular respiration (the process of breaking up covalent bonds within sugar molecules with the intake of oxygen and release of ATP, adenosine tri-phosphate). Mitochondria (plural of mitochondrion) are more numerous in cells requiring more energy (muscle, etc.). Mitochondria are self-replicating, containing their own DNA, RNA, and ribosomes. They have a double membrane; the internal membrane is folded. Cellular respiration reactions occur along the folds of the internal membrane (called **cristae**). Mitochondria are thought to be an evolved form of primitive bacteria (prokaryotic cells) that lived in a symbiotic relationship with eukaryotic cells more than 2 billion years ago. This concept, known as the **endosymbiont hypothesis**, is a plausible explanation of how mitochondria, which have many of the necessary components for life on their own, became an integral part of eukaryotic cells.

10. The **nucleus** is an organelle surrounded by two lipid bilayer membranes. The nucleus contains chromosomes, nuclear pores, nucleoplasm, and a nucleolus. The **nucleolus (10a)** is a rounded area within the nucleus of the cell where ribosomal RNA is synthesized. This rRNA is incorporated

into ribosomes after exiting the nucleus. Several nucleoli (plural of *nucleolus*) can exist within a nucleus. The **nuclear membrane (10b)** is the boundary between the nucleus and the cytoplasm. The nuclear membrane is actually a double membrane, which allows for the entrance and exit of certain molecules through the nuclear pores. **Nuclear pores (10c)** are points at which the double nuclear membrane fuses together, forming a passageway between the inside of the nucleus and the cytoplasm outside the nucleus. Nuclear pores allow the cell to selectively move molecules in and out of the nucleus. There are many pores scattered about the surface of the nuclear membrane.

Plant Cells

The structure of plant cells differs noticeably from animal cells with the addition of three organelles: the cell wall, the chloroplasts, and the central vacuole. In Figure 3-9, the organelles numbered 1 to 7 function the same way in plant cells as in animal cells (see above).

1. **Golgi apparatus**

2. **Mitochondria**

3. **Rough endoplasmic reticulum**

4. **Ribosome**

5. **Nucleus**

6. **Nucleolus**

7. **Smooth endoplasmic reticulum**

8. **Cell walls** surround plant cells. (Bacteria also have cell walls.) Cell walls are made up of cellulose and lignin, making them strong and rigid (whereas the cell membrane is relatively weak and flexible). The cell wall encloses the cell membrane, providing strength and protection for the cell. The cell wall allows plant cells to store water under relatively high concentration. The combined strength of a plant's cell walls provides support for the whole organism. Dry wood and cork are essentially the cell walls of dead plants. The structure of the cell wall allows substances to pass through it readily, so transport in and out of the cell is still regulated by the cell membrane.

9. The **cell membrane** (or plasma membrane) functions in plant and animal cells in the same way. However, in some plant tissues, channels connect the cytoplasm of adjacent cells.

10. **Chloroplasts** are found in plant cells (and also in some protists). Chloroplasts are the site of photosynthesis within plant cells. **Chlorophyll** pigment molecules give the chloroplast their green color, although the chloroplasts also contain yellow and red carotenoid pigments. In the fall, as chloroplasts lose chlorophyll, these pigments are revealed, giving leaves their red and yellow colors. The body (or **stroma**) of the chloroplast contains embedded stacked, disk-like plates (called **grana**), which are the site of photosynthetic reactions.

11. The **central vacuole** takes up much of the volume of plant cells. It is a membrane-bound fluid-filled space, which stores water and soluble nutrients for the plant's use. The tendency of the central vacuole to absorb water provides for the rigid shape (turgidity) of some plant cells. (Animal cells may also contain vacuoles for varying purposes, and these too are membrane-bound, fluid-filled spaces. For instance, contractile vacuoles perform the specific function of expelling waste and excess water from single-celled organisms.)

Fig. 3-9   A Typical Plant Cell.

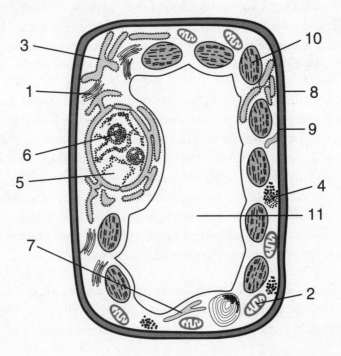

Enzymes

Enzymes are protein molecules that act as catalysts for organic reactions. A catalyst is a substance that lowers the activation energy of a reaction. A catalyst is not consumed in the reaction. Enzymes do not make reactions possible that would not otherwise occur under the right energy conditions, but they lower the activation energy, which increases the rate of the reaction.

Fig. 3-10   Effect of Enzyme on a Reaction. Adding an enzyme lowers the activation energy for a reaction.

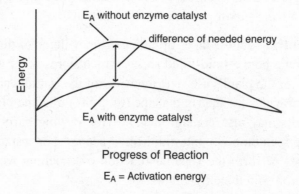

E_A without enzyme catalyst

difference of needed energy

Energy

E_A with enzyme catalyst

Progress of Reaction

E_A = Activation energy

Enzyme names end with the letters *-ase* and usually begin with a syllable describing the catalyzed substrate reaction (i.e., hydrolase catalyzes hydrolysis reactions; lactase catalyzes the breakdown of the sugar lactose). Thousands of reactions occur within cells, each controlled by one or more enzymes. Enzymes are synthesized within the cell at the ribosomes, as all proteins are.

Enzymes are effective catalysts because of their unique shapes. Each enzyme has a uniquely shaped area called its **active site**. For each enzyme, there is a particular substance known as its **substrate**, which fits within the active site (like a hand in a glove). When the substrate is seated in the active site, the combination of two molecules is called the **enzyme-substrate complex**. An enzyme can bind to two substrates and catalyze the formation of a new chemical bond, linking the two substrates. An enzyme may also bind to a single substrate and catalyze the breaking of a chemical bond, releasing two products. Once the reaction has taken place, the unchanged enzyme is released.

The function of enzymes lowers the energy needed to initiate cellular reactions. However, the completion of the reaction may either require or release energy. Remember from earlier in this chapter that reactions requiring energy are called endothermic reactions. Reactions that release energy are called exothermic reactions. Endothermic reactions can take place in a cell by being coupled to the breakdown of ATP or a similar molecule. Exothermic reactions are coupled to the production of ATP or another molecule with high-energy chemical bonds.

Fig. 3-11   Enzyme Reaction. 1: enzyme; 2: substrate; 3&4: enzyme-substrate complex; 5: products.

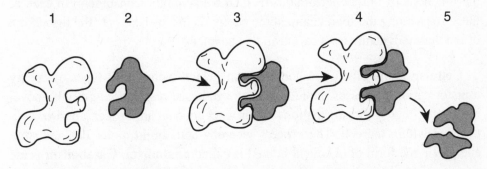

Some enzymatic reactions require a non-protein substance called a **cofactor**. The cofactor binds to the active site. This allows the substrate to fit into the active site. Some cofactors are inorganic. **Inorganic cofactors** include metal ions—for example, iron, copper, or zinc. Other cofactors are organic molecules. **Organic cofactors** are also called **coenzymes**. Some coenzymes are not made by cells but must be obtained in the diet. Most vitamins are coenzymes (or precursors of coenzymes). **Prosthetic groups** are similar to cofactors; they also facilitate the enzyme reaction. However, prosthetic groups are bound to the enzyme, rather than being separate atoms or molecules.

In some cases, other substances compete to attach to an enzyme's active site. If one of these substances, known as an **inhibitor,** attaches to the enzyme first, the cellular reaction will not take place. Environmental conditions within the cell, such as high temperature or acidity, may also inhibit an enzymatic reaction. These conditions may change the shape of the active site and render the enzyme ineffective (denaturing).

Enzyme reactions may also be controlled by mechanisms within the cell. Enzyme control (or **regulation**) may occur when the product of the reaction is also an inhibitor to the reaction. This slows down the production rate as the concentration of the product increases. In other cases, a particular molecule serves as a regulator, by changing the structure of the active site making the enzyme more or less effective.

Energy Transformations

All living things require energy. Ultimately, the source of most energy for life on Earth is the sun. Photosynthetic organisms (plants, some protists, and some bacteria) are able to harvest solar energy and transform it into chemical energy eventually stored within covalent bonds of molecules (such as carbohydrates, fats, and proteins). These organisms are called primary producers. Consumers eat producers and utilize the chemical energy stored in them to carry on the functions of life. Other organisms then consume the consumers. In each of these steps along the food chain, some energy is lost as heat (see the discussion of the thermodynamic laws earlier in this chapter).

Cellular metabolism is a general term, which includes all types of energy transformation processes, including photosynthesis, respiration, growth, movement, etc. Energy transformations occur as chemicals are broken apart or synthesized within the cell. The process whereby cells build molecules and store energy (in the form of chemical bonds) is called **anabolism**. **Catabolism** is the process of breaking down molecules and releasing stored energy.

ATP

ATP (adenosine triphosphate) is known as the energy currency of cellular activity. While energy is stored in the form of carbohydrates, fats, and proteins, the amount of energy contained within the bonds of any of these substances would overwhelm (and thus kill) a cell if released at once. In order for the energy to be released in small packets usable to a cell, large molecules need to be broken down in steps. ATP is an efficient storage molecule for the energy needed for cellular processes. ATP consists of a nitrogenous base (adenine), a simple sugar (ribose), and three phosphate groups. When a cellular process requires energy, a molecule of ATP can be broken down into ADP (adenosine diphosphate) plus a phosphate group. Even more energy is released when ATP is decomposed into AMP (adenosine monophosphate) plus two phosphate groups. Coupling these energy-releasing reactions with energy-absorbing reactions allows the cell to carry out its functions.

Photosynthesis

The process of **photosynthesis** includes a crucial set of reactions. These reactions convert the light energy of the sun into chemical energy usable by living things. Photosynthetic organisms use the converted energy for their own life processes, and also store energy that may be used by organisms that consume them.

Although the process of photosynthesis actually occurs through many small steps, the entire process can be summed up with the following equation:

$$6CO_2 + 6H_2O + \text{light energy} \rightarrow C_6H_{12}O_6 + 6O_2$$

(carbon dioxide + water $\rightarrow$ glucose + oxygen)

Chlorophyll is a green pigment (a pigment is a substance that absorbs light energy) that is able to absorb a photon of light, allowing photosynthesis to occur. Chlorophyll is contained in the grana of the chloroplast (see the discussion of plant cells earlier in this chapter). It is not consumed in the photosynthetic process, but must be present for the reactions to occur. There are two phases of the photosynthetic process: the light reactions, or photolysis, and the Calvin Cycle, or CO_2 fixation.

Energy from the sun is transformed by photosynthetic organisms into chemical energy in the form of ATP.

During the **light reactions (photolysis)**, the chlorophyll pigment absorbs a photon of light, leaving the chlorophyll in an excited (higher energy) state. The light reactions are decomposition reactions, which separate water molecules into hydrogen ions, oxygen ions, and high energy electrons. The oxygen atoms from the water combine to form O_2 (gas) and are released into the environment. The free hydrogens are grabbed and held by a particular molecule (called the hydrogen acceptor) until they are needed. The excited chlorophyll also supplies energy to a series of reactions that produce ATP from ADP and inorganic phosphate (P_i).

The **Calvin Cycle (CO_2 fixation)** then occurs in the stroma of the chloroplast. This second phase of photosynthesis does not require light; however, it does require the use of the products (hydrogen and ATP) of photolysis. In a multi-step process, six CO_2 molecules are linked with hydrogen (produced in photolysis) forming glucose (a six-carbon sugar). Glucose molecules can then be linked to form polysaccharides (starch or sugars), which are then stored in the cell.

Cellular Respiration

Unlike photosynthesis (which only occurs in photosynthetic cells), respiration occurs in all cells. Respiration breaks down molecules and releases energy for use by the cell. There are several steps involved in cellular respiration. Some require oxygen (that is, they are **aerobic** reactions) and some do not (they are **anaerobic**).

Glycolysis is the breaking down of glucose into smaller carbon-containing molecules; these breakdown reactions yield ATP (glyco = sugar, lysis = breakdown). It is the first step in all respiration pathways and occurs in the cytoplasm of all living cells. Each molecule of glucose (six carbons) is broken down into two molecules of pyruvic acid (or pyruvate) with three carbons each, two ATP molecules, and two hydrogen atoms (attached to NADH, nicotinamide adenine dinucleotide). This is an **anaerobic reaction** (no oxygen is required). The process of glycolysis is summarized by the following chemical equation:

$$\text{glucose (6 C)} + 2ADP + 2\ P_i + 2NAD^+ \rightarrow$$
$$\text{2 pyruvic acid (3 C each)} + 2ATP + 2NADH + 2H^+$$

After glycolysis, respiration will continue on one of two pathways, depending upon whether oxygen is present.

Aerobic Pathways

Aerobic respiration (in the presence of oxygen) begins with glycolysis and proceeds through two major steps, the **Krebs cycle** (also known as the citric acid cycle) and the **electron transport chain**. The first step, the Krebs cycle, occurs in the matrix of a cell's mitochondria and breaks down pyruvic acid molecules (three carbons each) into CO_2, H^+ (protons) and 2 ATP molecules. The Krebs cycle also liberates electrons, which then enter the next step.

The second step occurs along the **electron transport chain**, or ETC, which captures the energy released by the Krebs cycle. The ETC is a series of **cytochromes** on the cristae of the mitochondria. Cytochromes are pigment molecules, which include a protein and a **heme** (iron-containing) group. The iron in heme groups may be either oxidized (loses electron to form Fe^{+3}) or reduced (gains electron to form Fe^{+2}) as electrons are passed along the ETC. As electrons pass from one cytochrome to another, energy is released. Some of this energy is lost as heat; the rest is stored in molecules of ATP. This process can produce the most ATP per cycle, 32 ATP molecules per glucose molecule. The final step of the electron transport chain occurs when the last electron carrier transfers two electrons to an oxygen atom that simultaneously combines with two protons from the surrounding medium to produce water.

Anaerobic Pathways

If no oxygen is present within the cell, respiration will proceed anaerobically after glycolysis. Anaerobic respiration is also called **fermentation**. Anaerobic respiration breaks down the two pyruvic acid molecules (three carbons each) into end products (such as ethyl alcohol, C_2H_6O, and carbon dioxide (CO_2) or lactic acid $C_3H_6O_3$). The net gain from anaerobic respiration is two ATP molecules per glucose molecule. Fermentation is not as efficient as aerobic respiration; it uses only a small part of the energy available in a glucose molecule.

MOLECULAR BASIS OF HEREDITY

Watson and Crick were responsible for explaining the structure of the DNA molecule, which laid the foundation of our current understanding of the function of chromosomes and genes. Today, through the discoveries of these two scientists, and through the collaborative work of scientists worldwide, the study of chromosomes and genetic inheritance has proceeded to discover the intricacies of the **genomes** (sum total of genetic information) of many organisms, including humans. The study of genomes has developed further to include **genome mapping**, which allows fragments of DNA to be assigned to specific chromosomes. Maps are created for specific species based on results of studies of the genetic material found within that species.

A **gene** is a length of DNA that encodes for a particular protein. Each protein the cell synthesizes performs a specific function in the cell. The function of one protein, or the function of a group of proteins, is called a **trait**.

DNA Replication

In order to replicate, a portion of a DNA molecule unwinds, separating the two halves of the double helix. This separation is aided by the enzyme helicase. Another enzyme (DNA polymerase) binds to each strand and moves along them as it connects nucleotides using the original DNA strands as templates. The new strand is complementary to the original template and forms a new double helix with one of the parent strands. If no errors occur during DNA synthesis, the result is two identical double helix molecules of DNA.

The process of DNA replication, however, is occasionally subject to a mistake known as a **mutation**. All the DNA of every cell of every organism is copied repeatedly to form new cells for growth, repair, and reproduction. A mutation can result from an error that randomly occurs during replication. Mutations can also result from damage to DNA caused by exposure to certain chemicals, such as some solvents or the chemicals in cigarette smoke, or by radiation, such as ultraviolet radiation in sunlight or x-rays. Cells have built-in mechanisms for finding and repairing most DNA errors; however, they do not fix them all. The result of a DNA error, a mutation, in most cases expresses itself in a change (small or large) in the cell's structure and function.

DNA carries the information for making all the proteins a cell can make. The DNA information for making a particular protein can be called the gene for that protein. Genetic traits are expressed as a result of the activation of the combination of proteins encoded by the DNA of a cell. Protein synthesis occurs in two steps called transcription and translation.

Transcription refers to the formation of an RNA molecule, which corresponds to a gene. The DNA strand "unzips"; individual RNA nucleotides are strung together to match the DNA sequence by the enzyme RNA polymerase. The new RNA strand (known as messenger RNA or **mRNA**) then migrates from the nucleus to the cytoplasm (is modified in a process known as **post-transcriptional processing**). This processing prepares the mRNA for protein synthesis by removing the non-coding sequences. In the processed RNA, each unit of three nucleotides or **codon** encodes a particular amino acid.

The next phase of protein synthesis is called **translation**. In order for the protein synthesis process to continue, a second type of RNA is required, transfer RNA or **tRNA**. Transfer RNA is the link between the "language" of nucleotides (codon and anticodon) and the "language" of amino acids (hence the word "translation"). Transfer RNA is a chain of about 80 nucleotides. At one point along the tRNA chain, there are three unattached bases, which are called the anticodon. This anticodon will line up with a corresponding codon during translation. Each tRNA molecule also has an attached, specific amino acid.

Translation occurs at the ribosomes. A ribosome is a structure composed of proteins and ribosomal RNA (rRNA). A ribosome attaches to the mRNA strand at a particular codon known as the start codon. This codon is only recognized by a particular initiator tRNA. The ribosome continues to add tRNA whose anticodons make complementary bonds with the next codon on the mRNA string, forming a peptide bond between amino acids as each amino acid is held in place by a tRNA. At the end of the translation process, a terminating codon stops the synthesis process and the protein is released.

Structural and Regulatory Genes

Genes encode proteins of two varieties. **Structural genes** code proteins that form organs and structural characteristics. **Regulatory genes** code proteins that determine functional or physiological events, such as growth. These proteins regulate when other genes start or stop encoding proteins, which in turn produce specific traits.

Transduction and Transformation

In most organisms, DNA replication preserves a continuity of traits throughout the organism's life span. However, the genetic makeup of bacteria can be changed through one of two processes, transduction or transformation. **Transduction** is the transfer of genetic material (portions of a bacterial chromosome) from one bacterial cell to another. The transfer is mediated by a bacteriophage (a virus that targets bacteria). Bacteria may also absorb and incorporate pieces of DNA from their environment (usually from dead bacterial cells), in a process called **transformation**.

Cell Division

The process of cell reproduction is called **cell division** or mitosis. The process of cell division centers on the replication and separation of strands of **DNA**.

Structure of Chromosomes

Chromosomes are long chains of subunits called **nucleosomes**. Each nucleosome is composed of a short length of DNA wrapped around a core of small proteins called **histones**. The combination of DNA with histones is called **chromatin**. Each nucleosome is about 11 nm in diameter (a nanometer is one billionth of a meter) and contains a central core of eight histones with the DNA double helix wrapped around them. Each gene spans dozens of nucleosomes. The DNA plus histone strings are then tightly packed and coiled, forming chromatin.

In a cell that is getting ready to divide, each strand of chromatin is duplicated. The two identical strands (called **chromatids**) remain attached to each other at a point called the **centromere**. During cell division, the chromatin strands become more tightly coiled and packed, forming a chromosome, which is visible using a light microscope. At this stage, a chromosome consists of two identical chromatids, held together at the centromere, giving each chromosome an **X** shape.

Fig. 3-12   A Chromosome.

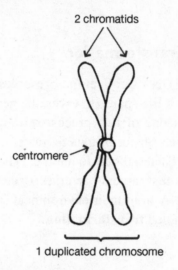

2 chromatids

centromere

1 duplicated chromosome

Within the nucleus, each chromosome pairs with another of similar size and shape. These pairs are called **homologs**. Each set of homologous chromosomes has a similar genetic constitution, but the genes are not necessarily identical. Different forms of corresponding genes are called **alleles**.

Fig. 3-13   Paired Homologous Chromosomes.

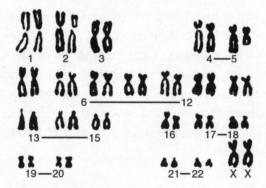

The Cell Cycle

A cell that is going to divide progresses through a particular sequence of events ending in cell division, which produces two daughter cells. This is known as the **cell cycle** (see Fig. 3-14). The time taken to progress through the cell cycle differs with different types of cells, but the sequence is the same. Cells in many tissues never divide (for example, cells in the central nervous system).

Fig. 3-14   The Cell Cycle. Interphase includes the G_1, S, and G_2 phases. The cell division phase includes mitosis and cytokinesis.

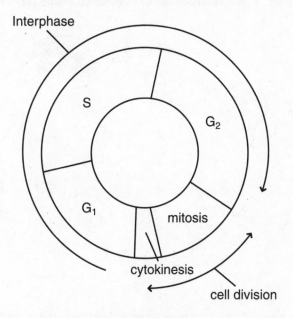

The Cell Cycle

There are three major periods within the cell cycle: interphase and mitosis (also called the M phase or cell division phase), and cytokinesis. **Interphase** is the period when the cell is active in carrying on its function. Interphase is divided into three phases. During the first phase, the **G_1 phase**, metabolism and protein synthesis are occurring at a high rate, and most of the growth of the cell occurs at this time. The cell organelles are produced (as necessary) and undergo growth during this phase. During the second phase, the **S phase**, the cell begins to prepare for cell division by replicating the DNA and proteins necessary to form a new set of chromosomes. In the final phase, the **G_2 phase**, more proteins are produced, which will be necessary for cell division, and the centrioles (which are integral to the division process) are replicated as well. Cell growth and function occur through all the stages of interphase.

Mitosis

Mitosis is the process by which a cell distributes its duplicated chromosomes so that each daughter cell has a full set of chromosomes and the nucleus divides. It is important to note one chromosome in each daughter cell is from each parent (maternal and paternal). In other words, mitosis is set up so that you don't end up with two maternal chromosome 5's in a single cell. Mitosis progresses through four phases: prophase, metaphase, anaphase, and telophase (see Fig. 3-15).

Fig. 3-15 Mitosis. See explanations of numbered steps below.

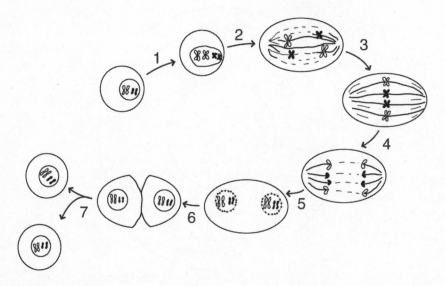

During **prophase** (**1, 2**), the first stage of mitosis, the chromatin condenses into chromosomes within the nucleus and becomes visible through a light microscope. The centrioles move to opposite ends of the cell, and **spindle fibers** begin to extend from each centriole toward the center of the cell. At this point, although the chromosomes become visible, the nucleolus no longer is. During the second part of prophase, the nuclear membrane dissolves and the spindle fibers attach to the centromeres, forming a junction called a **kinetochore**. The chromosomes then are pulled to the center by attached spindle fibers in preparation for the next step, metaphase.

During **metaphase** (**3**), the spindle fibers pull the chromosomes into alignment along the metaphase plate of the cell. This arrangement ensures that one copy of each chromosome is distributed to each daughter cell.

During **anaphase** (**4**), the chromatids are separated from each other when the centromere divides. Each former chromatid is now called a chromosome.

Each pair of identical chromosomes move along the spindle fibers to opposite ends of the cell.

Telophase (**5**) occurs as nuclear membranes form around the chromosomes. The chromosomes disperse through the new nucleoplasm and are no longer visible as chromosomes under a light microscope. The spindle fibers disappear. After telophase, the process of **cytokinesis** (**6**) produces two separate cells (**7**). Cytokinesis differs somewhat in plants and animals. In animal cells, a ring made up of the protein actin surrounds the center of the cell and contracts. As the actin ring contracts, it pinches the cytoplasm into two separate compartments. Each cell's plasma membrane seals, making two distinct daughter cells. In plant cells, a cell plate forms across the center of the cell and extends out towards the edges of the cell. When this plate reaches the edges, a cell wall forms on either side of the plate, and the original cell then splits into two.

Mitosis, then, produces two nearly identical daughter cells. Cells may differ in distribution of mitochondria or because of DNA replication errors, for example. Only eukaryotic unicellular organisms (protists and some fungi) reproduce by mitosis. Bacteria, which lack a nucelous, divide through **binary fission**.

Meiosis

Meiosis is the process of producing four daughter cells, each with a **haploid** set of unduplicated chromosomes. The parent cell is **diploid**, that is, it has a normal set of paired chromosomes. Meiosis goes through a two-stage process resulting in four new cells, rather than two (as in mitosis). Each cell has half the chromosomes of the parent. Meiosis occurs in reproductive organs, and the resultant four haploid cells are called **gametes** (egg and sperm). When two haploid gametes fuse during the process of fertilization, the resultant cell has one chromosome set from each parent and is diploid. This process allows for the huge genetic diversity among species.

Fig. 3-16 Meiosis. See explanations of numbered steps below.

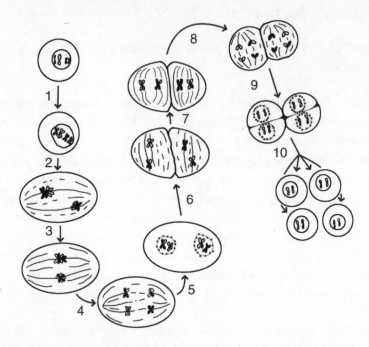

Two distinct nuclear divisions occur during meiosis, or reduction division (meiosis 1, steps **1 to 5** in Fig. 3-16) and division (or meiosis 2, steps **6 to 10**). **Reduction** affects the **ploidy** (referring to haploid or diploid) level, reducing it from 2n to n (i.e., diploid to haploid). **Division** then distributes the remaining set of chromosomes in a mitosis-like process.

The phases of meiosis 1 are similar to the phases of mitosis, with some notable differences. As in mitosis, chromosome replication (**1**) occurs before prophase; then during prophase 1 (**2**), homologous chromosomes pair up and join at a point called a **synapse** (this happens only in meiosis). The attached chromosomes are now termed a **tetrad**, a dense four-stranded structure composed of the four chromatids from the original chromosomes. At this point, some portions of the chromatid may break off and reattach to another chromatid in the tetrad. This process, known as **crossing over**, results in an even wider array of final genetic possibilities. Genetic engineering, discussed further in Chapter 6, involves manipulation of the genes of an organism, usually to obtain a desired characteristic.

The nuclear membrane disappears during late prophase (or prometaphase). Each chromosome (rather than each chromatid) develops a kinetochore, and as the spindle fibers attach to each chromosome, they begin to move.

In metaphase 1 (**3**), the two chromosomes (a total of four chromatids per pair) align themselves along the metaphase plate of the cell. Each homologous pair of chromosomes contains one chromosome from the mother and one from the father from the original sexual production of that organism. When the homologous pairs orient at the cell's center in preparation for separating, the chromosomes randomly sort. The resulting cells from this meiotic division will have a mixture of chromosomes from each parent. This increases the possibilities for variety among descendent cells.

Anaphase 1 (**4**) occurs next as one chromosome from each pair moves to separate ends of the cell. This phase differs from the anaphase of mitosis where one of each chromosome pair (rather than one chromatid) separates. In telophase 1 (**5**), the nuclear envelope may or may not form, depending on the type of organism. The cell divides and two daughter cells are formed. The cell then proceeds to meiosis 2.

The nuclear envelopes dissolve (if they have formed) during prophase 2 (**6**) and spindle fibers form again. All else proceeds as in mitosis, through metaphase 2 (**7**), anaphase 2 (**8**), and telophase 2 (**9**). Again, as in mitosis, each chromosome splits into two chromatids. The process ends with cytokinesis (**10**), forming four distinct gamete cells.

Restriction Enzymes

The study of DNA has been greatly aided by the discovery of restriction enzymes (restriction endonucleases). Restriction enzymes cut sections of DNA molecules by cleaving the sugar-phosphate backbone at a particular nucleotide sequence.

Scientists have isolated hundreds of different restriction endonucleases that act on a few hundred different DNA sequences. Restriction enzymes are made by bacteria and act to destroy foreign DNA (for example, viral DNA) that has entered the bacterial cell.

In the laboratory, restriction enzymes are used to cut DNA into small strands of DNA for study.

Restriction enzymes are generally named after their host of origin, rather than the substrate upon which they act. For example, EcoRI is from the host *Escherichia coli*, Hind II and Hind III from *Haemophilus influenzae*, XhoI from *Xanthomonas holcicola*, etc.

CHAPTER 4

Plants (Botany)

PLANTS (BOTANY)

Most of us commonly recognize plants as organisms that produce their own food through the process of photosynthesis. (Some bacteria are also photosynthetic.) However, the plant kingdom is divided into several classifications according to physical characteristics.

Vascular plants (tracheophytes) have tissue organized in such a way as to conduct food and water throughout their structure. These plants include some that produce seeds (such as corn or roses) as well as those that do not produce any seeds (such as ferns). **Nonvascular** plants (bryophytes), such as mosses, lack specialized tissue for conducting water or food. They produce no seeds or flowers and are generally only a few centimeters in height.

Another method of classifying plants is according to their method of reproduction. **Angiosperms** are plants that produce flowers, which are reproductive organs. **Gymnosperms**, on the other hand, produce seeds without flowers. These include conifers (cone-bearers) and cycads.

Plants that survive only through a single growing season are known as **annuals**. Other plants are **biennial**; their life cycle spans two growing seasons. **Perennial** plants continue to grow year after year.

PLANT ANATOMY

Plants have structures that allow them to thrive in their environment. Angiosperms and gymnosperms differ mostly in the structure of their stems and reproductive organs. Gymnosperms are mostly trees and shrubs, with woody, instead of herbaceous stems. Gymnosperms do not produce flowers; instead they produce seeds in cones or cone-like structures.

Fig. 4-1 A Typical Flowering Plant (angiosperm). Note descriptions of numbered structures in text.

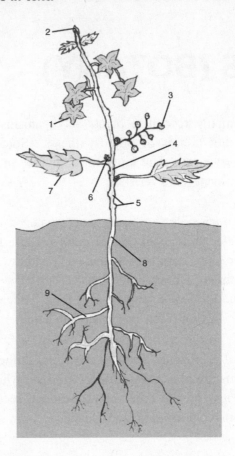

Angiosperms

The shoot system of angiosperms includes the stem, leaves, flowers, and fruit, as well as growth structures such as nodes and buds (see Fig. 4-1). The signature structure of an angiosperm is the **flower (1)**, the primary reproductive organ. Before the flower blooms, it is enclosed within the **sepals (1-a)**, small, green, leaf-like structures, which fold back to reveal the flower **petals (1-b)**. The petals usually are brightly colored; their main function is to attract insects and birds, which may be necessary for the process of pollination. The short branch or stem, which supports the flower, is called the **pedicel (1-c)**. (See Fig. 4-2.)

Usually (but depending on the species), a single flower will have both male and female reproductive organs. The **pistil** is the female structure, and includes the stigma, style, ovary, and ovules. The **stigma (1-d)** is a sticky surface at the top of the pistil, which traps pollen grains. The stigma sits above a slender vase-like structure, the **style (1-e)**, which connects to the ovary. The **ovary (1-f)** is the hollow, bulb-shaped structure in the lower interior of the pistil. After seeds have formed, the ovary will ripen and become fruit. Within the ovary are the **ovules (1-g)**, small round cases each containing one or more egg cells. If the egg is fertilized, the ovule will become a seed. In the process of meiosis in the ovule, an egg cell is produced, along with smaller bodies known as polar nuclei. The polar nuclei will develop into the endosperm of the seed when fertilized by sperm cells.

Fig. 4-2   Typical Flower. Note numbered explanations in text.

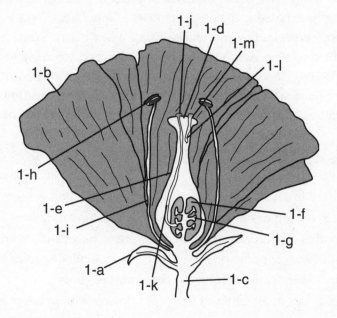

The male structure is the stamen, consisting of the **anther (1-h)** atop the long, hollow **filament (1-i)**. The anther has four lobes and contains cells (microspore mother cells) that become pollen. Some mature **pollen grains (1-j)** are conveyed (usually by wind, birds, or insects) to a flower of a compatible species, where they stick to the stigma. The stigma produces chemicals, which stimulate the pollen to burrow into the style, forming a hollow **pollen tube (1-k)**. This tube is produced by the tube **nucleus (1-l)**, which has developed from a portion of the pollen grain. The pollen tube extends down toward the ovary. Behind the tube nucleus are two **sperm nuclei (1-m)**. When the sperm nuclei reach the ovule, one will join with an egg cell, fertilizing it to become a zygote (the beginning cell of the embryo). The other sperm nucleus merges with the polar bodies forming the endosperm, which will feed the growing embryo.

The **shoot apex (2)** is composed of **meristem** tissue (consisting of undifferentiated cells capable of quick growth and specialization) and is the region where elongation of the stem occurs. The **terminal bud** (the beginning of a new set of leaves) is also located at the shoot apex. Each year, as the plant continues to grow taller, a new terminal bud and shoot apex are produced. The spot where the previous year's terminal bud was located is then called a **terminal bud scar**.

Fruit (3) is a matured ovary, which contains the seeds (mature fertilized ovules). The fruit provides protection for the seeds, as well as a method to disburse them. For instance, when ripened fruit is eaten by animals, the seeds are discarded or excreted in the animal's waste, transferring the seed to a new location for germination. Each **seed** contains a tiny embryonic plant, stored food, and a seed coat for protection. When the seed is exposed to the proper amount of moisture, temperature, and oxygen, it germinates (begins to sprout and grow into a new plant). Stored food in a seed is found in the cotyledon. Angiosperms are also classified according to the structure of their cotyledons. Plants with two cotyledons in each seed are known as dicotyledons (**dicots**); those with only one are known as monocotyledons (**monocots**). The following chart outlines the major differences between dicots and monocots:

Dicots	Monocots
e.g., oaks, flowers, vegetables	e.g., grasses, lilies, palm trees
two cotyledons in seed	one cotyledon in seed
leaves have branched or networked veins	leaves have parallel veins
vascular bundles (collections of xylem and phloem tubes) arranged in rings	stems have random arrangement of vascular bundles
taproot system with smaller secondary roots	fibrous roots
flowers with petals in multiples of four or five	flowers with petals in multiples of three

The **stem (4)** is the main support structure of the plant. The stem produces leaves and lateral (parallel with the ground) branches. **Nodes (5)** are the locations along the stem where new leaves sprout, and the space between nodes is the **internode**. New leaves begin as **lateral buds (6)**, which can be seen on growing plants.

The stem is also the main organ for transporting food and water to and from the leaves. In some cases the stem also stores food; for instance, a potato is a tuber (stem) that stores starch. The stem also contains meristem tissue.

Most of the stem tissue is made up of **vascular tissue**, including two varieties—**xylem** and **phloem**. Xylem tissue is composed of long tubular cells, which transport water up from the ground to the branches and leaves. Phloem tissue, formed by stacked cells connected by sieve plates which allow nutrients to pass from cell to cell, transports food made in the leaves (by photosynthesis) to the rest of the plant.

The **leaf (7)** is the primary site of photosynthesis in most plants. Most leaves are thin, flat, and joined to a branch or stem by a petiole (a small stem-like extension). The petiole houses vascular tissue, which connects the veins in the leaf with those in the stem.

The **cuticle**, which maintains the leaf's moisture balance, covers most leaf surfaces. Consider a cross-section of a leaf (see Fig. 4-3), the outermost layer is the **epidermis (7-a, e)**. The epidermis is generally one cell thick. It secretes the waxy cuticle and protects the inner tissue of the leaf.

The mesophyll is composed of several layers of tissue between the upper and lower epidermis. The uppermost, the **palisade layer (7-b)**, contains vertically aligned cells with numerous chloroplasts. The arrangement of these cells maximizes the potential for exposure of the chloroplasts to needed sunlight. Most photosynthesis occurs in this layer.

The sugars produced by photosynthesis are transported throughout the plant via the **vascular bundles (7-c)** of xylem and phloem. The vascular bundles make up the veins in the leaf.

The next layer beneath the palisade cells is the **spongy layer (7-d)**, a layer of parenchyma cells separated by large air spaces. The air spaces allow for the exchange of gases (carbon dioxide and oxygen) for photosynthesis.

On the underside of the leaf, there are openings ringed by **guard cells (7-f)**. The openings are called **stomata (7-g)** (or stomates). The stomata serve to allow moisture and gases (carbon dioxide and oxygen) to pass in and out of the leaf, thus facilitating photosynthesis.

Fig. 4-3   Cross-section of a leaf.

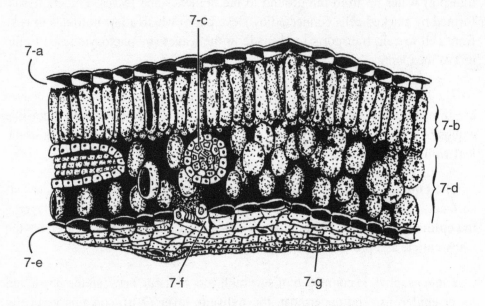

The root system of a typical angiosperm includes the **primary roots (8)**, which extend downward, and the **lateral roots (9)**, which develop secondarily and extend horizontally, parallel with the ground surface. Roots function to provide water and needed nutrients to the plant. Roots are structured to provide a large surface area for absorption. The network of the root system also anchors the plant.

Roots have four major structural regions, which run vertically from bottom to top. The **root cap** is composed of dead, thick-walled cells and covers the tip of the root, protecting it as the root pushes through soil. The **meristematic region** is just above the root cap. It consists of undifferentiated cells, which carry on mitosis, producing cells that grow to form the **elongation region**. In the elongation region, cells differentiate, large vacuoles are formed, and cells grow. As the cells differentiate into various root tissues, they become part of the **maturation region**.

A cross-section of root tissue above the maturation region would reveal several types of **primary root tissue**. In the maturation region, the epidermis produces **root hairs (9-a)**, extensions of the cells, which reach between soil particles and retrieve water and minerals. The primary tissues include the outermost layer, the **epidermis (9-b)**. The epidermis is one cell layer thick and serves to protect the internal root tissue and absorb nutrients and water.

Fig. 4-4   Root cross-section.

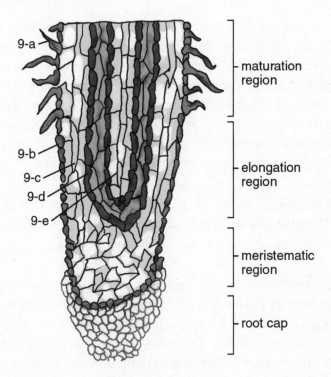

Inside the epidermis is a ring known as the **cortex (9-c)**, which is made up of large parenchyma cells. **Parenchyma** cells are present in many tissues of plants; they are thin-walled cells loosely packed to allow for flow of gases and uptake of minerals.

Inside the cortex is a ring of **endodermis (9-d)**, a single layer of cells, which are tightly connected so no substances can pass between cells. This feature allows the endodermis to act as a filter; all substances entering the vascular tissues from the root must pass through these cells. In the center of the root is the **vascular cylinder (9-e)**, including xylem and phloem tissue.

PLANT PHYSIOLOGY

Water and Mineral Absorption and Transport

Although plants produce their own sugars and starches for food, they must obtain water, carbon dioxide, and minerals from their environment. Vascular plants have well-developed systems for absorption and transport of water and minerals.

Water is essential to all cells of all plants, so plants must have the ability to obtain water and transport water molecules throughout their structure. Most water is absorbed through the plant's root system, then makes its way in one of two pathways toward the xylem cells, which will transport water up the stem and to the leaves and flowers. The first pathway is for water to seep between the epidermal cells of the roots and between the parenchyma cells of the cortex. When water reaches the endodermal tissue, it enters the cells and is pushed through the vascular tissue toward the xylem by root pressure.

A second pathway is for the water to pass through the cell wall and plasma membrane. Water travels along this intracellular route through channels in the cell membranes (plasmodesmata), until it reaches the xylem.

Once water reaches the xylem, hydrogen bonding between water molecules (known as **cohesion**) causes tension that pulls water through the water column up through the stem and onto the leaves (known as the **cohesion-tension process**). Some water that has traveled up through the plant to the leaves is evaporated, a process known as **transpiration**. As water is evaporated, it causes a siphoning effect (like sucking on a straw), which continues to pull water up from the root xylem, through the length of the plant, and to the leaves.

Food Translocation and Storage

Food is manufactured by photosynthesis mostly in the leaves. The rest of the plant must have this food (carbohydrates) imported from the leaves. The leaves have source cells, which manufacture sugars. The food molecules are transferred from the source cells to phloem tissue through active transport (energy is expended to move molecules across the plasma membrane against the concentration gradient—from low concentration to high concentration). Once in the phloem, the sugars begin to build up, causing osmosis to occur (water enters the phloem lowering the sugar concentration). The entrance of water into the phloem causes pressure, which pushes the water-sugar solution through **sieve plates** that join the cells. This pressure thrusts the water-sugar solution to all areas of the plant, making food available to all cells in the plant.

C3 and C4 Plants

It is important to know that there are differences in the physiology of different plants. Accounting for over 95% of species of plants on the Earth are C3 plants. In photosynthesis, these plants use the enzyme rubisco to make a three-carbon compound during the process of carbon fixation. The best climate for C3 plants would be cool, damp, and cloudy. Loss of carbon through photorespiration is high. However, the metabolic process for these plants is more energy efficient, and, therefore, requires lower light levels.

C4 plants, which include many grasses such as sugar cane and maize, have an extremely low rate of carbon loss through photorespiration. The best habitat for these plants is hot and dry environments. C4 plants use water very efficiently.

PLANT REPRODUCTION AND DEVELOPMENT

The reproductive cycle of plants occurs through the alternation of haploid (n) and diploid (2n) phases. [Remember from Chapter 3 that the haploid cells have one complete set of chromosomes (n). Diploid cells have two sets of chromosomes (2n).] Diploid and haploid stages are both capable of undergoing mitosis in plants. The diploid generation is known as a **sporophyte**. The reproductive organs of the sporophyte produce **gametophytes** through the process of meiosis. Gametophytes may be male or female and are haploid. The male gametophyte produces **sperm (male gamete)**; the female produces an **egg cell (female gamete)**. When a sperm cell **fertilizes** an egg cell (haploid cells join to form a diploid cell), they produce a **zygote**. The zygote will grow into an **embryo**, which resides within the growing seed.

We are accustomed to identifying particular plants according to their adult phase, which is only one phase of the life cycle. Various phyla of plants have their own identifiable life cycles, which include an **alternation of generations**.

Fig. 4-5 Alternation of Generations in Ferns.

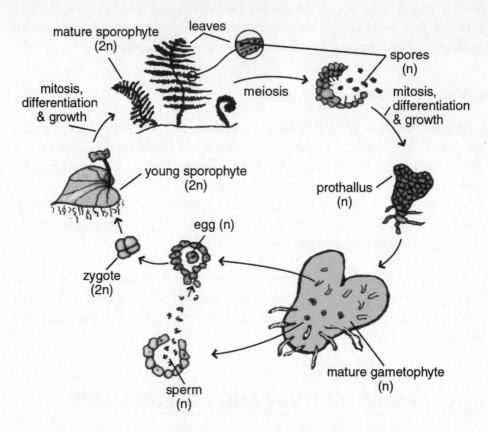

Mosses and ferns alternate haploid and diploid phases, developing two distinct generations of the plant, each with its own recognizable form. One generation is haploid, the other diploid. The haploid phase is most prominent in mosses; while in ferns, the diploid stage is most prominent.

Fig. 4-6 **Pine Life Cycle. The adult tree produces both male (pollen) and female (ovulate) cones that form the pollen and ovules that combine to produce a seed.**

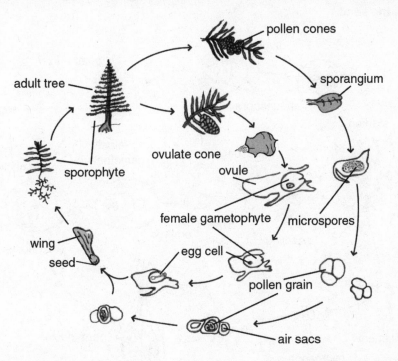

In conifers (such as pines), the sporophyte generation (diploid) is the familiar adult of the species. The process of meiosis produces the haploid gametophytes (male and female) from the male and female cones. The male gametophyte forms the male pollen grain and its attached air bladders, which assist it in being borne by the wind. The pollen contains sperm cells and tube cells, which will fertilize an egg cell of a female scale when they are brought into contact.

In angiosperms, the dominant adult generation is also the sporophyte—the flowering plant. (See Fig. 4-7.)

Asexual Plant Reproduction

Some plants may also reproduce through **vegetative propagation**—an asexual process. Asexual reproduction occurs through mitosis only (it does not involve gametes), and produces offspring genetically identical to the parent. While sexual reproduction leads to genetic variation and adaptation, asexual reproduction of a plant with a desirable set of genetic traits, preserves these intact in successive generations. Many plants reproduce through a combination of sexual and asexual reproduction, reaping the advantages of each.

Fig 4-7 Angiosperm Life Cycle. Pollen develops from the stamen; eggs develop from the pistil. The pollen fertilizes the egg, resulting in an embryo within the seed.

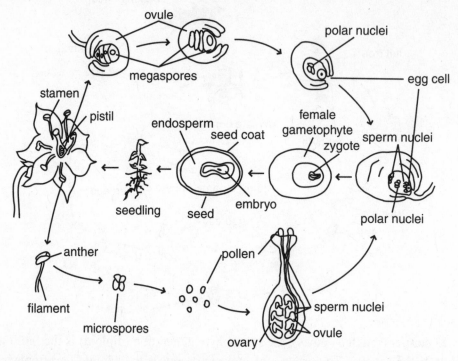

There are several types of plants that produce structures specifically designed to carry on vegetative propagation. These are described in the chart below:

Reproductive Structure	Description of Structure	Plants with these Structures
tubers	underground storage stems, develop new shoots after dormant season	potatoes
rhizomes	underground runners that develop into new plant	irises
stolens	above-ground runners that grow roots of their own then develop into new plant	strawberries
bulbs	underground storage units that grow into many new plants via division	amaryllis
corms	resemble bulbs but with enlarged, solid stem for food storage	gladiolus, crocus

PLANT GROWTH AND DEVELOPMENT

Hormones are chemicals that regulate the growth, development, and function of an organism. Plant cells produce hormones that bring about physiological changes within plant tissues. Each type of hormone affects changes in particular cells known as target cells. The traits regulated by hormones are many and varied, but there are some particularly important ones in plants. Note the most common hormones and their functions in the following chart:

Hormone	Process Regulated or Influenced
giberellins	cell division and cell elongation (65 hormones)
cytokinins	cell division and fruit development
abscisic acid	opening and closing of stomata (controlling water lost through transpiration and formation of winter buds that put plant in dormant state)
ethylene	ripening of fruit (spoiling releases ethylene which stimulates ripening of surrounding fruit); metabolic activity (i.e., producing female flowers to increase fertilization)
auxins	growth factors (i.e., tropisms)

A **tropism** is an involuntary response of an organism to an external stimulus such as light, water, gravity, or nutrients. For instance, plant stems are usually positively **phototropic** (they grow towards light); while plant roots are negatively phototropic (they grow away from light). Plant roots are positively **geotropic**; they grow toward the center of the earth, while stems are negatively geotropic, growing against gravity. Tropisms are thought to be caused by plant hormones, which react to the external stimulus causing some cells to grow quickly and others to grow slowly. Variations of auxin levels also influence the strength of petioles and stems, regulating when leaves or fruit drop.

There are other factors (besides hormones) that influence plant growth and development. For instance, plants respond to relative periods of light and darkness, a characteristic known as **photoperiodicity**. Light-sensitive chemicals in the leaves trigger a response in the plant, which encourage growth, flowering, or other reactions. Photoperiodicity causes flowering and growth of varying plants at different times of year.

CHAPTER 5

Animals (Zoology)

CHAPTER 5

ANIMALS (ZOOLOGY)

The animal kingdom includes a wide variety of phyla that have a range of body plans. This range includes certain invertebrates with relatively simple body plans as well as highly complex vertebrates (including humans). There are specific characteristics that differentiate animals from other living things. Organisms in the animal kingdom share the following traits:

1. Animal cells do not have cell walls or plastids.

2. Adult animals are multi-cellular with specialized tissues and organs.

3. Animals are heterotrophic (they do not produce their own food).

4. Animal species are capable of sexual reproduction, although some are also capable of asexual reproduction (e.g., hydra).

5. Animals develop from embryonic stages.

In addition to the above traits, most adult animals have a symmetrical anatomy. Adult animals can have either radial symmetry (constituent parts are arranged radiating symmetrically about a center point) or bilateral symmetry (the body can be divided along a center plane into equal, mirror-image halves). There are a few exceptions to this rule, including the adult sponge whose body is not necessarily symmetrical. While there is wide variation in the physical structure of animals, the animal kingdom is usually divided into two broad categories—invertebrates and vertebrates. There are many more species of invertebrates than vertebrates.

Invertebrates are those species having no internal backbone structure; **vertebrates** have internal backbones. Invertebrates include sponges and worms, which have no skeletal structure at all, and arthropods, mollusks, crustaceans, etc., which have exoskeletons. In fact, there are many more phyla of invertebrates than vertebrates (about 950,000 phyla of invertebrates and only about 40,000 phyla of vertebrates).

ANIMAL ANATOMY

Tissues

Like all multicellular organisms, animal bodies contain several kinds of tissues, made up of different cell types. Differentiated cells may be organized into specialized tissues performing particular functions. There are eight major types of animal tissue:

1. **Epithelial tissue** consists of thin layers of cells. Epithelial tissue makes up the layers of skin, lines ducts and the intestine, and covers the inside of the body cavity. Epithelial tissue forms the barrier between the environment and the interior of the body.

2. **Connective tissue** covers internal organs and composes ligaments and tendons. This tissue holds tissues and organs together, stabilizing the body structure.

3. **Muscle tissue** is divided into three types—smooth, skeletal, and cardiac. **Smooth** muscle makes up the walls of internal organs and functions in involuntary movement (breathing, digestion, etc.). **Skeletal** muscle attaches bones of the skeleton to each other and surrounding tissues. Skeletal muscle's function is to enable voluntary movement. **Cardiac** muscle is the tissue forming the walls of the heart. Its strength and electrical properties are vital to the heart's ability to pump blood.

4. **Bone tissue** is found in the skeleton and provides support, protection for internal organs, and ability to move as muscles pull against bones.

5. **Cartilage tissue** reduces friction between bones, and supports and connects them. For example, it is found at the ends of bones, and in the ears and nose.

6. **Adipose tissue** is found beneath the skin and around organs, providing cushioning, insulation, and fat storage.

7. **Nerve tissue** is found in the brain, spinal cord, nerves, and ganglia. It carries electrical and chemical impulses to and from organs and limbs to the brain. Nerve tissue in the brain receives these impulses and sustains mental activity.

8. **Blood tissue** consists of several cell types in a fluid called plasma. It flows through the blood vessels and heart, and is essential for carrying

oxygen to cells, fighting infection, and carrying nutrients and wastes to and from cells. Blood also has clotting capabilities, which preserve the body's functions in case of injury.

Tissues are organized into organs, and organs function together to form systems, which support the life of an organism. Studying these systems allows us to understand how organisms thrive within their ecosystem.

Systems

Many different body plans exist amongst animals, and each type of body plan includes systems necessary for the organism to live. Our discussion of systems here focuses on those found in most vertebrates. Vertebrates are highly complex organisms with several systems working together to perform the functions necessary to life. These include the digestive, respiratory, skeletal, nervous, circulatory, excretory, and immune systems.

Digestive System

The **digestive system** (see Fig. 5-1) serves as a processing plant for ingested food. The digestive system in animals generally encompasses the processes of **ingestion** (food intake), **digestion** (breaking down of ingested particles into molecules that can be absorbed by the body), and **egestion** (the elimination of indigestible materials). In most vertebrates, the digestive organs are divided into two categories, the **alimentary canal** and the **accessory organs**. The alimentary canal is also known as the **gastrointestinal** (or GI) **tract** and includes the mouth, pharynx, esophagus, stomach, small intestine, large intestine, rectum, and anus. The accessory organs include the teeth, tongue, salivary glands, liver, gallbladder, and pancreas.

The **mouth** (oral cavity, **1**) is the organ of ingestion and the first organ of digestion in the GI tract. The first step in digestion in many vertebrates occurs as food is chewed. Chewing is the initial step in breaking down food into particles of manageable size. Chewing also increases the surface area of the food and mixes it with saliva, which contains the starch-digesting enzyme amylase. Saliva is secreted by the **salivary glands** (**2**). Chewed food is then swallowed and moved toward the **stomach** (**3**) by peristalsis (muscle contraction) of the **esophagus** (**4**). The stomach is a muscular organ that stores incompletely digested food. The stomach continues the mechanical and chemical breakdown of food particles begun by the chewing process. The lining of the stomach secretes mucous to protect it from the strong digestive chemicals necessary in the digestive process. The stomach also secretes digestive enzymes and hydrochloric acid, which continue the digestive process to the point of producing a watery soup of nutrients, called chyme, which then proceeds through the pyloric sphincter into the small intestine (the duodenum). The **pancreas (5)** and **gall bladder (6)** release more enzymes into the small intestine, the site where the final steps of digestion and most absorption occurs. The cells lining the **small intestine (7)** have protrusions out into the lumen of the intestine called **villi**. Villi provide a large surface area for absorption of nutrients. Nutrients move into the capillaries through or between the cells making up the villi. The enriched blood travels to the **liver (8)**, where some sugars are removed and stored. The indigestible food proceeds from the small intestine to the **large intestine (9)** where water is absorbed back into the body. The waste **(feces)** is then passed through the **rectum (10)** and excreted through the **anus (11)**.

Fig. 5-1 Human Digestive System.

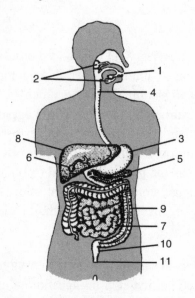

There are, of course, variations in the digestive system among animals of various classifications. Some vertebrates, such as cows and deer, are **ruminants**; they consume large amounts of vegetation. These animals have several chambers in their stomachs. Chewed vegetation is regurgitated from the first two stomach chambers as **cud**, and is chewed again, allowing much of their food to be broken down mechanically. Bacteria in the digestive track then break down cellulose, the main constituent of a ruminant's diet.

Many invertebrates, such as insects and earthworms, have digestive systems resembling those of vertebrates, including a mouth, esophagus, stomach, and intestines. Many of these species also have a **crop**, an organ that stores food until it is processed for absorption. Other animals have only a sac-like digestive cavity that performs the necessary functions of digestion.

Respiratory or Gas Exchange System

Also known as the **respiratory system** (see Fig. 5-2), the **gas exchange system** is responsible for the intake and processing of gases required by an organism, and for expelling gases produced as waste products. In humans, air is taken in primarily through the **nose** (although gases may be inhaled through the mouth, the nose is better at filtering out pollutants in the air). The **nasal passages (1)** have a mucous lining to capture foreign particles. This lining is surrounded by epithelial tissue with embedded capillaries, which serve to warm the entering air. Air then passes through the **pharynx (2)** and into the **trachea (3)**. The trachea includes the windpipe or **larynx** in its upper portion and the **glottis**, an opening allowing gases to pass into the two branches known as the bronchi. The glottis is guarded by a flap of tissue, the **epiglottis**, which prevents food particles from entering the bronchial tubes. The **bronchi (4)** lead to the two **lungs (5)** where they branch out in all directions into smaller tubules known as **bronchioles (6)**.

Fig. 5-2   Human Respiratory System.

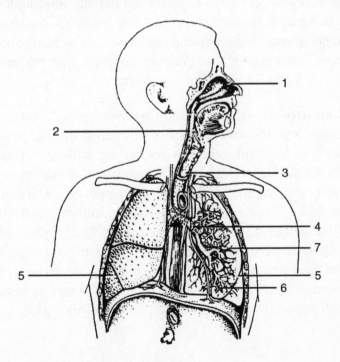

The bronchioles end in **alveoli (7)**, thin-walled air sacs, which are the site of gas exchange. The bronchioles are surrounded by capillaries, which bring blood with a high density of carbon dioxide and a low concentration of oxygen from the pulmonary arteries. At the alveoli, the carbon dioxide diffuses from the blood into the alveoli and oxygen diffuses from the alveoli into the blood. The oxygenated blood is carried away to tissues throughout the body.

All living organisms require the ability to exchange gases, and there are several variations to the means and organs utilized for this life process. Invertebrates, such as the earthworm, are able to absorb gases through their skin. Insects rely on the diffusion of gases through holes in the exoskeleton known as spiracles. In single-celled organisms such as the amoeba, diffusion of gases occurs directly through the plasma membrane.

Musculoskeletal System

The **musculoskeletal system** provides the body with structure, stability, and the ability to move. The musculoskeletal system is unique to vertebrates, although some invertebrates (such as mollusks and insects) have external support structures (exoskeletons) and muscle.

In humans, the musculoskeletal system is composed of joints, ligaments, cartilage, muscle groups, and 206 bones. The skeleton provides protection for the soft internal organs, as well as structure and stability allowing for an upright stature and movement. Bones also perform the important function of storing calcium and phosphates, and producing red blood cells within the bone marrow. The 206 bones forming the human skeleton are linked with movable joints and joined by muscle systems controlling movement.

Skeletal muscles are voluntary—they are activated by command from the nervous system. **Smooth muscle** lines most internal organs, protecting their contents and function, and generally contracting without conscious intent. For instance, the involuntary (automatic) contraction of smooth muscle in the esophagus and lungs facilitates digestion and respiration. **Cardiac muscle** is unique to the heart. It is involuntary muscle (like smooth muscle), but cardiac muscle also has unique features, which cause it to "beat" rhythmically. Cardiac muscle cells have branched endings that interlock with each other, keeping the muscle fibers from ripping apart during their strong contractions. In addition, electrical impulses travel in waves from cell to cell in cardiac muscle causing the muscle to contract in a coordinated way with a rhythmic pace.

Nervous System

The **nervous system** is a communication network that connects the entire body of an organism and provides control over bodily functions and actions. Nerve tissue is composed of nerve cells known as **neurons** and **glial cells**. Neurons carry impulses via electrochemical responses through their **cell body** and **axon** (long root-like appendage of the cell). Nerve cells exist in networks, with axons of neighbor neurons interacting across small spaces (**synapses**). Chemical neurotransmitters send messages along the nerve network causing responses specific to varying types of nerve tissue. The nervous system allows the body to sense stimuli and conditions in the environment and respond with necessary reactions. **Sensory organs**—skin, eyes, nose, ears, etc.—transmit signals in response to environmental stimuli to the **brain**. The brain then conveys messages via nerves to glands and muscles, which produce the necessary response.

Fig. 5-3 A Typical Neuron.

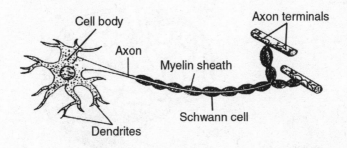

The human nervous system (and that of many mammals) is anatomically divided into two systems, the central nervous system and the peripheral nervous system. The following outline shows the components of each portion of the nervous system.

I. **Central Nervous System** (CNS)—two main components are the **brain** and the **spinal cord**. These organs control all other organs and systems of the body. The spinal cord is a continuation of the brain stem and acts as a conduit of nerve messages.

II. **Peripheral Nervous System** (PNS)—a network of nerves throughout the body.

  A. **Sensory Division**

    1. **visceral sensory nerves**—carry impulses from body organs to CNS

2. **somatic sensory nerves**—carry impulses from the body surface to CNS

B. **Motor Division**

1. **somatic motor nerves**—carries impulses to skeletal muscle from CNS

2. **autonomic nervous system**

a. **sympathetic** nervous system—carries impulses that stimulate organs

b. **parasympathetic** nervous system—carries impulses back from organs

Fig. 5-4 Human Nervous System.

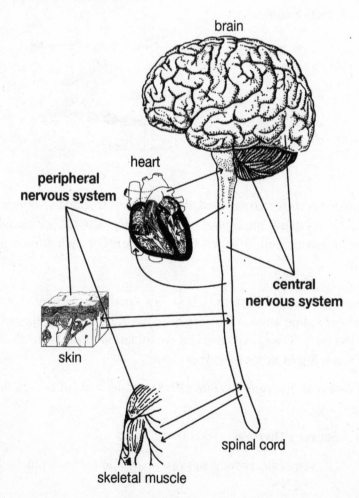

The brain of vertebrates has three major divisions: the forebrain, midbrain, and hindbrain. The **forebrain** is located most anterior and contains the **olfactory lobes** (sense smell), **cerebrum** (controls sensory and motor responses, memory, speech, and most factors of intelligence), as well as the **thalamus** (integrates senses), **hypothalamus** (is involved in hunger, thirst, blood pressure, body temperature, hostility, pain, pleasure, etc.), and **pituitary gland** (releases various hormones). The **midbrain** is between the forebrain and hindbrain and contains the **optic lobes** (visual center connected to the eyes by the optic nerves). The **hindbrain** consists of the **cerebellum** (controls balance, equilibrium, and muscle coordination) and the **medulla oblongata** (controls involuntary response such as breathing and heartbeat).

Within the brain, nerve tissue is grayish in color and is called **gray matter**. The nerve cells, which exist in the spinal cord and throughout the body, have insulation covering their axons. This insulation (called the **myelin sheath)** speeds electrochemical conduction within the axon of the nerve cell. Since the myelin sheath gives this tissue a white color, it is called **white matter**. The myelin sheath is made up of individual cells called Schwann cells.

The nervous systems of vertebrates and some invertebrates are highly sophisticated, providing conscious response and unconscious controls. However, the nervous systems of some species of invertebrates (such as jellyfish) are relatively simple networks of neurons that control only some aspects of their body functions.

Circulatory System

The process of cellular metabolism is a fundamental process of life and cannot proceed without a continuous supply of oxygen to every living cell within the body. The **circulatory system** is the conduit for delivering nutrients and gases to all cells and for removing waste products from them.

In invertebrates, the circulatory system may consist entirely of diffusion in the gastrovascular cavity, or it may be an **open circulatory system** (where blood directly bathes the internal organs) or a **closed circulatory system** (where blood is confined to vessels).

Closed circulatory systems are more typical of vertebrates. In vertebrates, **blood** flows throughout the circulatory system within **vessels**. Vessels include **arteries**, **veins**, and **capillaries**. The pumping action of the **heart** (a hollow, muscular organ) forces blood in one direction throughout the system. In large animals, valves within the heart, and some of the vessels in limbs, keep blood from flowing backwards (being pulled downward by gravity).

Blood carries many products to cells throughout the body, including minerals, infection-fighting white blood cells, nutrients, proteins, hormones, and metabolites. Blood also carries dissolved gases (particularly oxygen) to cells and waste gases (mainly carbon dioxide) away from cells.

Capillaries (tiny vessels) surround all tissues of the body and exchange carbon dioxide for oxygen. Oxygen is carried by **hemoglobin** (containing iron) in red blood cells. Oxygen enters the blood in the lungs and travels to the heart, then through **arteries** (larger vessels that carry blood away from the heart) and **arterioles** (small arteries), to capillaries. The blood picks up carbon dioxide waste from the cells and carries it through capillaries, then **venules** (small veins), and **veins** (vessels that carry blood toward the heart) back to the heart and on to the lungs. Thus, blood is continually cycled.

Fig. 5-5   Human Circulatory System. Blood flows from the heart through arteries to the capillaries throughout the body and returns via the veins.

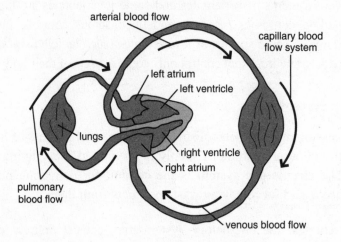

Excretory System

The **excretory system** is responsible for collecting waste materials, filtering waste out of body fluids, and transporting them to organs that expel them from the body. There are many types of waste that must be expelled from the body, and there are many organs involved in this process.

The primary excretory organs of most vertebrates are the kidneys. The **kidneys** filter metabolic wastes from the blood and excrete them as **urine** into the urinary tract. The urinary tract carries the fluid that is eventually expelled from the body. Urine is typically 95% water, and may contain urea (formed from breakdown of proteins), uric acid (formed from breaking down nucleic acids), creatinine (a byproduct of muscle contraction), and various minerals and hormones.

The **skin** is an accessory excretory organ; salts, urea, and other wastes are secreted with water from sweat glands in the skin.

The **liver** produces **bile** which aids in digesting fats and also carries away broken down pigments and chemicals (often from pollutants and medications) and secretes them into the small intestine where they proceed to the large intestine and are expelled in the feces. The liver also breaks down some nitrogenous molecules (including some proteins) then excreting them as urea. The **lungs** are the sites of excretion for carbon dioxide.

Immune System

The **immune system** functions to defend the body from infection by bacteria and viruses. The **lymphatic system** is the principal infection-fighting component of the immune system. The organs of the lymphatic system in humans and other higher invertebrates include the lymph, lymph nodes, spleen, thymus, and tonsils. **Lymph** is a collection of excess fluid that is absorbed from between cells into a special system of vessels, which circulates through the lymphatic system and finally empties into the bloodstream. Lymph also collects plasma proteins that have leaked into interstitial fluids.

Lymph nodes are small masses of lymph tissue whose function is to filter lymph and produce lymphocytes. **Lymphocytes** and other cells are involved in the immune system. Lymphocytes begin in bone marrow as stem cells and are collected and distributed via the lymph nodes. There are two classes of lymphocytes, B cells and T cells. **B cells** emerge from the bone marrow and produce **antibodies**, which enter the bloodstream. These antibodies find and attach themselves to foreign **antigens** (toxins, bacteria, foreign cells, etc.). The attachment of an antibody to an antigen marks the pair for destruction.

The **spleen** contains some lymphatic tissue and is located in the abdomen. It filters larger volumes of lymph. The **tonsils** are a group of lymph cells connected together and located in the throat.

The **thymus** is another mass of lymph tissue, which is active only through the teen years, fighting infection and producing T cells. **T cells** mature in the thymus gland. Some T cells (like B cells) patrol the blood for antigens, but T cells are also equipped to destroy antigens themselves. T cells also regulate the body's immune responses.

Homeostatic Mechanisms

All living cells, tissues, organs, and organisms must maintain a tight range of physical and chemical conditions in order for them to live. Conditions such as temperature, pH, water balance, sugar levels, etc., must be monitored and controlled in order to keep them within the accepted ranges that will not inhibit life. When the conditions of an organism are within acceptable ranges, it is said to be in **homeostasis**. Organisms have a special set of mechanisms that serve to keep them in homeostasis. Homeostasis is a state of dynamic equilibrium, which balances forces tending toward change and forces acceptable for life functions.

There are many instances of feedback control. These take effect when any situation arises that may drive levels out of the normal acceptable range. In other words, the homeostatic mechanism is a reaction to a stimulus. This reaction, called a **feedback response**, is the production of some counterforce that levels the system.

Homeostasis is achieved mostly by actions of the sympathetic and parasympathetic nervous systems by a process known as **feedback control**. For instance, when the body undergoes physical activity, muscle action causes a rise in temperature. If not checked, rising temperature can destroy cells. In this instance, the nervous system detects the rising temperature and reacts with a response that causes sweat glands to produce sweat. The evaporation of sweat cools the body.

Hormonal Control in Homeostasis and Reproduction

Hormones are chemicals produced in the endocrine glands of an organism, which generally travel through the circulatory system and are taken up by specific targeted organs or tissues, where they modify metabolic activities.

A hormone is manufactured in response to a particular **stimulus**. The hormone (for instance a **steroid**) enters the bloodstream from one of the ductless endocrine glands that manufacture hormones. The steroid passes through the cell membrane of the targeted cell and enters the cytoplasm. The hormone combines with a particular protein known as a receptor, creating the **hormone-receptor complex**. This complex enters the nucleus and binds to a DNA molecule causing a gene to be transcribed. The mRNA molecule leaves the nucleus for the endoplasmic reticulum, where it encodes a particular protein. The protein migrates to the site of the stimulus and counteracts the source of the stimulus.

The second process targets receptors on a cell's membrane. A particular **receptor** exists on the membrane when the cell is in a particular condition (for instance containing an excess of glucose). When the hormone binds with the receptor on the membrane, the receptor changes its form. This triggers a chain of events within the cytoplasm resulting in the production or destruction of proteins thus moderating the conditions.

Hormones control many physiological functions, from digestion to conscious responses and thinking and to reproduction. In humans, for instance, women of childbearing age have a continuous cycle of hormones. The hormone cycle causes the release of eggs at specific times. If the egg is fertilized, a different combination of hormones stimulates a chain of events that promotes the development of the embryo.

ANIMAL REPRODUCTION AND DEVELOPMENT

Reproduction in multicellular animals is a complex process that generally proceeds through the steps of **gametogenesis** (gamete formation) and then **fertilization**.

Gametes are the sex cells—sperm and eggs formed in the reproductive organs. When a sperm of one individual combines with the egg cell of another, the resulting cell is known as a **zygote**. A **zygote** then develops into a new individual. In the case of **spermatogenesis** (sperm formation), diploid **primary spermatocytes** are formed from special cells (**spermatogonia**) in the testes. The primary spermatocytes then undergo meiosis I, forming haploid **secondary spermatocytes** with a single chromosome set. (Please see the section on meiosis in Chapter 3.) The secondary spermatocytes go through meiosis II, forming **spermatids**, which are haploid. These spermatids then develop into the **sperm cells**.

In human female reproductive organs, egg cells are formed through a similar process known as **oogenesis**. **Primary oocytes** are typically present in great number in the female's ovaries at birth. Primary oocytes undergo meiosis I forming one **secondary oocyte** and one smaller **polar body**. Both the secondary oocyte and the polar body undergo meiosis II; the polar body producing two polar bodies (not functional cells), and the oocyte producing one more polar body and one haploid **egg cell**. The egg cell is now ready for fertilization, and if there are sperm cells present, the egg may be fertilized forming a diploid cell with a new combination of chromosomes, the zygote.

All multicellular organisms that reproduce sexually begin life as a zygote. The zygote then undergoes a series of cell divisions known as **cleavage**. After the first few divisions, the cluster of cells is called a morula. The **morula** then continues cell division, and the cluster begins to take shape as a thin layer of cells surrounding an internal cavity, the **blastula**. As cell division continues, the cells migrate and rearrange themselves, transforming the blastula into a two-layered cup shape, called the **gastrula** (a process known as **gastrulation**). As the gastrula develops, the cup shape reforms itself into a double-layered tube. The ectoderm, mesoderm, and endoderm form through the process of gastrulation and are collectively called the **germ layers**. As the germ layers develop, the embryo becomes recognizable, and differentiation continues until the organ systems are fully developed.

The outer layer of the gastrula tube will become the **ectoderm**, which later will develop into the skin, some endocrine glands, and the nervous system. The inner layer of the tube will become the **endoderm**, the precursor of the gut lining and various accessory structures. With further development, a third layer, between the ectoderm and endoderm arises—the **mesoderm**. The mesoderm layer will eventually form muscles and organs of the skeletal, circulatory, respiratory, reproductive, and excretory systems.

Fig. 5-6   Human Extraembryonic Membranes.

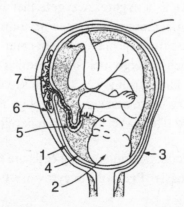

In addition to forming the tissues and organ systems of vertebrates, the germ layers also develop into **extraembryonic membranes** (i.e., membranes not part of the embryos themselves; see Fig. 5-6). The first of these membranes is the **chorion (1)**. In egg-laying vertebrates, the chorion lies in contact with the innermost surface of the shell, while in other vertebrates it is the outermost membrane surrounding the **embryo (2)** and in contact with the **uterus (3)**. In both cases, the chorion functions in regulating the passage of gases and water from the embryo to its surrounding environment. In embryos without shells, the chorion also controls passage of nutrients and wastes between the embryo and the mother.

Within the chorion is the **amnion (4)**, a fluid-filled (**amniotic fluid**) sac enclosing the embryo. The amniotic fluid cushions the embryo and helps keep temperatures constant. The fluid also keeps the amnionic membrane from sticking to the developing embryo.

The third membrane is the **allantois (5)**. It arises from the developing digestive tract. In humans and other vertebrates that bear live young, the allantois appears in the third week of development and becomes part of the **umbilical cord**. It contains blood vessels, which function to exchange gases and nutrients between the embryo and the mother. In egg-laying reptiles, the allantois is a reservoir for wastes. It fuses with the chorion, forming the **chorioallantoic membrane**, which regulates gas exchanges through the shell.

The **yolk sac membrane (6)**, enclosing the **yolk sac (7)**, also forms from the developing digestive tract and also becomes part of the umbilical cord. The yolk sac stores nutrients for use by the embryo. The yolk sac is larger and contains more material in egg-laying species, since there is no continuing contact with the mother. The yolk sac cells also give rise to gametes, which develop in reproductive organs of the embryo.

In mammals, the outer cells of the embryo and the inner cells of the uterus combine to form the **placenta**. The placenta is the connection between the mother and embryo; it is the site of transfer for nutrients, water, and wastes between them. The embryo synthesizes its own blood that is kept separate from the mother's blood. In the placenta, the vessels (that connect the circulatory system of the embryo through the umbilical cord to the placenta) pass right next to the mother's blood vessels. Nutrients, water, and oxygen diffuse from the mother's blood to the embryo's, while wastes and carbon dioxide diffuse into the mother's blood supply.

CHAPTER 6

Principles of Heredity (Genetics)

CHAPTER 6

PRINCIPLES OF HEREDITY (GENETICS)

The process by which characteristics pass from one generation to another is known as **inheritance**. The study of the principles of heredity (now called genetics) advanced greatly through the experimental work of **Gregor Mendel** (c. 1865). Mendel studied the relationships between traits expressed in parents and offspring, and the hereditary factors that caused expression of traits.

Mendel systematically bred pea plants to determine how certain hereditary traits passed from generation to generation. First, he established true-breeding plants, which produce offspring with the same traits as the parents. For example, the seeds of pea plants with yellow seeds would grow into plants that produced yellow seeds. Green seeds grow into plants that produce green seeds. Mendel named this first generation of true-breeding plants the parent or P_1 **generation**; he then bred the plant with yellow seeds and the plant with green seeds. Mendel called the first generation of offspring the F_1 **generation**. The F_1 generation of Mendel's yellow seed/green seed crosses contained only yellow seed offspring.

Mendel continued his experiment by crossing two individuals of the F_1 generation to produce an F_2 **generation**. In this generation, he found that some of the plants (one out of four) produced green seeds. Mendel performed hundreds of such crosses, studying some 10,000 pea plants, and was able to establish the rules of inheritance from them. The following are Mendel's main discoveries:

- Parents transmit hereditary factors (now called **genes**) to offspring. Genes then produce a characteristic, such as seed-coat color.

- Each individual carries two copies of a gene, and the copies may differ.

- The two genes an individual carries act independently, and the effect of one may mask the effect of the other. Mendel coined the terms geneticists still use: "dominant" and "recessive."

MODERN GENETICS

We now know that **chromosomes** carry all the genetic information in most organisms. Most organisms have corresponding pairs of chromosomes that carry genes for the same traits. These pairs are known as **homologous chromosomes**. Genes that produce a given trait exist at the same position (or **locus**) on homologous chromosomes. Each gene may have different forms, known as **alleles**. For instance, yellow seeds and green seeds arise from different alleles of the same gene. A gene can have two or more alleles, which differ in their nucleotide sequence. That difference can translate into proteins that function differently resulting in variations of the trait.

Sexual reproduction (meiosis) produces gamete cells with ½ the genetic information of the parents (paired chromosomes are separated and sorted independently). Therefore, each gamete may receive one of any number of combinations of each parent's chromosomes.

In addition, a trait may arise from one or more genes. However, because one-gene traits are easiest to understand, we will use them for most of our examples. If a trait is produced from a gene or genes with varying alleles, several possibilities for traits exist. The combination of alleles that make a particular trait is the **genotype**, while the trait expressed is the **phenotype**.

An allele is considered **dominant** if it masks the effect of its partner allele. The allele that does not produce its trait when present with a dominant allele is **recessive**. That is, when a dominant allele pairs with a recessive allele, the expressed trait is that of the dominant allele.

A **Punnett square** is a notation that allows us to easily predict the results of a genetic cross. In a Punnett square, a letter is assigned to each gene. Uppercase letters represent dominant traits, while lowercase letters represent recessive traits (a convention begun by Mendel). The possible alleles from each parent are noted across the top and side of a box diagram; then the possible offspring are represented within the internal boxes. If we assign the allele that produces yellow seeds the letter **Y**, and the allele that produces green seeds **y**, we can represent Mendel's first cross between pea plants (**YY** × **yy**) by the following Punnett square:

	Y	Y
y	Yy	Yy
y	Yy	Yy

One parent pea plant had green seeds (green seeds is its phenotype), so it must not have had any of the dominant genes for yellow seeds (**Y**). Therefore, it must have the genotype **yy**. If the second parent had one allele for yellow and one for green, then some of the offspring would have inherited two genes for green. Since Mendel started with true-breeding plants, we may deduce that one parent had two genes for green seeds (**yy**) and the other had two genes for yellow seeds (**YY**).

When both alleles for a given gene are the same in an individual (such as **YY** or **yy**), that individual is **homozygous** for that trait. Furthermore, the individual's genotype can be called homozygous. Both of the above parents (**P₁**) were homozygous. The children in the **F₁** generation all have one dominant gene (**Y**) and one recessive gene (**y**); their phenotype is yellow; and their genotype is **Yy**. When the two alleles for a given gene are different in an individual (**Yy**), that individual is said to be **heterozygous** for that trait; its genotype is heterozygous.

Breeding two **F₁** offspring from the example above produces the following Punnett square of a double heterozygous (both parents **Yy**) cross:

	Y	y
Y	YY	Yy
y	Yy	yy

Through this Punnett square, we can determine that three-fourths of the offspring will produce yellow seeds. This is consistent with Mendel's findings. However, there are two different genotypes represented among the yellow seed offspring. One-half of the offspring were heterozygous yellow (**Yy**), and one-fourth were homozygous yellow (**YY**).

The example above shows a **monohybrid cross**—a cross between two individuals where only one trait is considered. Mendel also experimented with crossing two parents while considering two separate traits, a **dihybrid cross**.

The laws investigated by Mendel form the basis of modern genetics. However, Mendel's laws now incorporate modern terminology (i.e., "genes" rather than "hereditary factors," etc.).

The Law of Segregation

The first law of Mendelian genetics is the **law of segregation**. The law of segregation states that traits are expressed from a pair of genes in the individual (on homologous chromosomes). Each parent provides one chromosome of every pair of homologous chromosomes. Paired chromosomes (and thus corresponding genes) separate and randomly recombine during gamete formation.

The Law of Dominance

Mendel determined that one gene usually expressed itself over the other (was dominant). This is the **law of dominance**, Mendel's second law of inheritance. In Mendel's experiments, the first generation produced no plants with green seeds, leading him to recognize the existence of genetic dominance. The yellow-seed allele was clearly dominant.

The Law of Independent Assortment

Mendel also investigated whether genes for one trait always were linked to genes for another. In other words, Mendel experimented not only with pea seed-coat color, but also with pea-plant height (and a number of other traits in peas and other plants). He wanted to determine if the parent plant that had green seeds and was tall would produce all plants with green seeds and be tall. These dihybrid cross experiments demonstrated that most traits were independent of one another. That is, a pea plant could be green and tall or green and short, yellow and tall or yellow and short. In most cases, genes for traits randomly sort into pairs (although some genes lie close to others on a chromosome and therefore can be inherited together). Since homologous chromosomes separate and independently sort in gamete formation, alleles are also separated and independently sorted, an assertion known as the **law of independent assortment**.

The following Punnett square demonstrates independent assortment. **Y** stands for the allele for yellow color, **y** for the allele for green, **T** for the allele tall, and **t** for short:

	TY	Ty	tY	ty
TY	TTYY	TTYy	TtYY	TtYy
Ty	TTYy	TTyy	TtYy	Ttyy
tY	TtYY	TtYy	ttYY	ttYy
ty	TtYy	Ttyy	ttYy	ttyy

Incomplete Dominance

Some traits are determined by genes that are neither dominant nor recessive and instead produce offspring that are a mix of the two parents. For instance, in snapdragons a plant with red flowers crossed with a plant with white flowers produces offspring with pink flowers. This is known as **incomplete dominance**. The dominant gene is not fully expressed. Neither white nor red is dominant over the other. In incomplete dominance, the conventional way to symbolize the alleles is with a capital letter designating the trait (in this case C for color) and a superscript designating the allele choices (in this case ^{R} for red, ^{W} for white), making the possible alleles C^R and C^W. The following Punnett square represents the incomplete dominance of the allele for red flowers (C^R), the allele for white (C^W), and the combination resulting in pink (C^RC^W).

	C^R	C^R
C^W	C^RC^W	C^RC^W
C^W	C^RC^W	C^RC^W

In this case, two plants, one with white flowers and one with red, are crossed to form all pink flowers. If two of the heterozygous offspring of this cross are then bred, the outcome of this cross ($C^RC^W \times C^RC^W$) will be:

	C^R	C^W
C^R	C^RC^R	C^RC^W
C^W	C^RC^W	C^WC^W

One-fourth of the offspring will be red, one-half pink, and one-fourth white, a 1:2:1 ratio.

Multiple Alleles

In the instances above, two possible alleles exist in a species, so the genotype will be a combination of those two alleles. There are some instances where more than two choices of alleles are present. For instance, for human blood types there is a dominant allele for type A blood, another dominant allele for type B blood, as well as a recessive allele, i, for neither A nor B, known as type O blood. There are three different alleles and they may combine in any way. In multiple-allele crosses, it is conventional to denote the chromosome by a letter (in this case **I** for dominant, **i** for recessive), with a subscript letter representing the allele types (in this case **A**, **B**, or **O**). The alleles for A and B blood are co-dominant, while the allele for O blood is recessive. The possible genotypes and phenotypes then are as follows:

genotype	phenotype
$I^A I^A$	Type A blood
$I^B I^B$	Type B blood
$I^B i^O$	Type B blood
$I^A i^O$	Type A blood
$I^A I^B$	Type AB blood
$i^O i^O$	Type O blood

[Note: There is another gene responsible for the Rh factor that adds the + or − to the blood type.]

Linkage

While Mendel had established the law of independent assortment, later study of genetics by other scientists found that this law was not always true. In studying fruit flies, for instance, it was found that some traits are always inherited together; they were not independently sorted. Traits that are inherited together are said to be **linked**. Genes are portions of chromosomes, so most traits produced by genes that are close together on the same chromosome are inherited together. The chromosomes are independently sorted, not the individual genes.

However, an exception to this rule complicates the issue. During metaphase of meiosis I, when homologous chromosomes line up along the center of the dividing cell, some pieces of the chromosomes break off and move from one chromosome to another (change places). This random breaking and reforming of homologous chromosomes allows genes to change the chromosome they are linked to, thus changing the genome of that chromosome. This process, known as **crossing over**, adds even more possibility of variation of traits among species. It is more likely for crossing over to occur between genes that do not lie close together on a chromosome than those that lie close together.

Gender is determined in an organism by a particular homologous pair of chromosomes. The symbols **X** and **Y** denote the sex chromosomes. In mammals and many insects, the male has an **X** and **Y** chromosome (**XY**), while the female has two **X**'s (**XX**). Genes that are located on the gender chromosome **Y** will only be seen in males. It would be considered a **sex-limited trait.** An example of a sex-limited trait is bar coloring in chickens that occurs only in males.

Some traits are **sex-linked**. In sex-linked traits, more males (**XY**) develop the trait because males have only one copy of the **X** chromosome. Females have a second **X** gene, which may carry a gene coding for a functional protein for the trait in question that may counteract a recessive trait. These traits (for example, hemophilia and colorblindness) occur much more often in males than females.

Still other traits may be **sex-influenced**. In this case, the trait is known as autosomal—it only requires one recessive gene to be expressed if there is no counteracting dominant gene. A male with one recessive allele will develop the trait, whereas a female would require two recessive genes to develop it. An example of a sex-influenced trait is male-pattern baldness.

Polygenic Inheritance

While the best-studied genetic traits arise from alleles of a single gene, most traits, such as height and skin color, are produced from the expression of more than one set of genes. Traits produced from interaction of multiple sets of genes are known as **polygenic traits**. Diseases such as diabetes and heart disease have multiple contributing factors involved in their development in a person. Besides genetic factors, these diseases can be all or partly caused by environmental or lifestyle factors.

Polygenic traits are difficult to map and difficult to predict because of the varied effects of the different genes and contributing factors on a specific trait or disease.

CHAPTER 7

Population Biology

POPULATION BIOLOGY

ECOLOGY

Ecology is the study of how organisms interact with other organisms and how they influence or are influenced by their physical **environment**. The word "ecology" is derived from the Greek term *oikos* (meaning "home" or "place to live") and *ology* (meaning "the study of"), so ecology is a study of organisms in their home. This study has revealed a number of patterns and principles that help us understand how organisms relate to their environment. First, however, it is important to grasp some basic vocabulary used in ecology.

The study of ecology centers on the ecosystem. An **ecosystem** is a group of populations found within a given locality, including the abiotic environment around those populations. A **population** is the total number of a single species of organism found in a given ecosystem. Typically, there are many populations of different species within a particular ecosystem. The term **organism** refers to an individual of a particular species. Each species is a distinct group of individuals that are able to interbreed (mate), producing viable offspring. Although species are defined by their ability to reproduce, they are usually described by their morphology (their anatomical features).

Populations that interact with each other in a particular ecosystem are collectively termed a **community**. For instance, a temperate forest community includes pine trees, oaks, shrubs, lichen, mosses, ferns, squirrels, deer, insects, owls, bacteria, fungi, etc.

The part of the Earth that includes all living things is called the **biosphere**. The biosphere also includes the **atmosphere** (air), the **lithosphere** (ground), and the **hydrosphere** (water).

A **habitat** refers to the physical place where a species lives. A species' habitat must include all the factors that will support its life and reproduction. These factors may be **biotic** (i.e., living—food source, predators, etc.) and **abiotic** (i.e., nonliving—weather, temperature, soil features, etc.).

A species' **niche** is the role it plays within the ecosystem. It includes its physical requirements (such as light and water) and its biological activities (how it reproduces, how it acquires food, etc.). One important aspect of a species' niche is its place in the food chain.

Ecological Cycles

Every species within an ecosystem requires resources and energy in varying forms. The interaction of organisms and the environment can be described as cycles of energy and resources that allow the community to flourish. Although each ecosystem has its own energy and nutrient cycles, these cycles also interact with each other to form bioregional and planetary biological cycles.

The **energy cycle** supports life throughout the environment. There are also several **biogeochemical cycles** (the water cycle, the carbon cycle, the nitrogen cycle, the phosphorous cycle, etc.), which are also important to the health of ecosystems. A biogeochemical cycle is the system whereby the substances needed for life are recycled and transported throughout the environment.

Carbon, hydrogen, oxygen, phosphorous, and nitrogen are called macronutrients; they are used in large quantities by living things. Micronutrients, those elements utilized in trace quantities in organisms, include iodine, iron, zinc, and copper.

Energy Cycle (Food Chain)

Since all life requires the input of energy, the **energy cycles** within the ecosystem are central to its well-being. On Earth, the Sun provides the energy that is the basis of life in most ecosystems. An exception is the hydrothermal vent communities that derive their energy from the heat of the Earth's core. Without the constant influx of solar energy into our planetary ecosystem, most life would cease to exist. Energy generally flows through the entire ecosystem in one direction—from producers to consumers and on to decomposers (consumers may also consume decomposers) through the **food chain**.

Photosynthetic organisms—such as plants, some protists, and some bacteria—are the first link in most food chains; they use the energy of sunlight to combine carbon dioxide and water into sugars, releasing oxygen gas (O_2). Photosynthetic organisms are called producers, since they synthesize sugar and starch molecules using the Sun's energy to link the carbons in carbon dioxide. Primary consumers (also known as herbivores) are species that eat photosynthetic organisms. Consumers utilize sugars and starches stored in cells or tissues for energy. Secondary consumers feed on primary consumers, and on the chain goes, through tertiary, and quarternary consumers. Finally, decomposers (bacteria, fungi, some animals) are species that recycle the organic material found in dead plants and animals back into the food chain.

Animals that feed only on other animals are called **carnivores** (meat-eaters), whereas those that consume both photosynthetic organisms and other animals are known as **omnivores**.

The energy cycle of the food chain is subject to the laws of thermodynamics. Energy can neither be created nor destroyed. However, every use of energy is less than 100% efficient; about 10% is lost as heat. When we call photosynthetic organisms producers, we mean that they produce food, using the Sun's energy to form chemical bonds in sugars and other biomolecules. Other organisms can use the energy stored in the bonds of these biomolecules.

The steps in the food chain are also known as **trophic levels**. Consider the pyramid diagram (Fig. 7-1) as one example of a food chain with many trophic levels. Grasses are on the bottom of the pyramid; they are the producers, the first trophic level. Producers are also known as **autotrophs**, as they produce their own food. Each trophic level decreases in **biomass** (total mass of organisms) from the level below it.

Fig. 7-1 A Food Chain Pyramid.

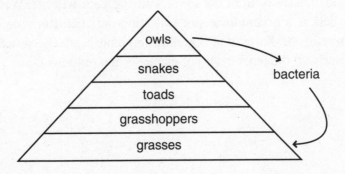

Grasshoppers represent the **second trophic level,** or primary consumers in this example of a food chain. Grasshoppers consume plants and are consumed (in this example) by toads, the secondary consumers, which represent the **third trophic level**. Snakes consume toads and are in turn consumed by owls—making these the **fourth** and **fifth trophic levels**. In this example, bacteria are the decomposers that recycle some of the nutrients from dead owls (and other levels) to be reused by the first trophic level.

The pyramid illustrates a food chain; however, in nature it is never actually as simple as shown. Owls consume snakes, but they may also consume toads (a lower level in the pyramid) and fish (from an entirely different pyramid). Thus, within every ecosystem there may be numerous food chains interacting in varying ways to form a **food web**. Furthermore, all organisms produce waste products that feed decomposers. The food web represents the cycling and recycling of both energy and nutrients within the ecosystem. The productivity of the entire web is dependent upon the amount of photosynthesis carried out by producers.

Water Cycle

The availability of water is crucial to the survival of all living things. Water vapor circulates through the biosphere in a process called the **hydrologic cycle**. Water is evaporated via solar radiation from the ocean and other bodies of water into clouds. Water is also released into the atmosphere from vegetation (leaves) by transpiration. Some water is also evaporated directly from soil, but most water in the ground flows into underground aquifers, which eventually empty into the oceans. Water above ground flows into waterways, which also eventually flow into the ocean (a process known as runoff). Water vapor is then redistributed over land (and back into oceans as well) via clouds, which release water as precipitation.

The water cycle also has a profound effect on Earth's climate. Clouds reflect the Sun's radiation away from the Earth, causing cool weather. Water vapor in the air also acts as a **greenhouse gas**, reflecting radiation from the Earth's surface back toward the Earth, and therefore trapping heat. The water cycle also intersects nearly all the other cycles of elements and nutrients.

Nitrogen Cycle

Nitrogen is another substance essential to life processes, since it is a key component of amino acids (components of proteins) and nucleic acids. The nitrogen cycle recycles nitrogen. Nitrogen is the most plentiful gas in the atmosphere, making up 78% of the air. However, neither photosynthetic organisms nor animals are able to use nitrogen gas (N_2), which does not readily react with other compounds, directly from the air. Instead, a process known as nitrogen fixation makes nitrogen available for absorption by the roots of plants. **Nitrogen fixing** is the process of combining nitrogen with either hydrogen or oxygen, mostly by **nitrogen-fixing bacteria**, or to a small degree by the action of **lightning**.

Nitrogen-fixing bacteria live in the soil and perform the task of combining gaseous nitrogen from the atmosphere with hydrogen forming ammonium (NH_4^+) ions. Some cyanobacteria, also called blue-green bacteria, are also active in this process. Ammonium ions are then absorbed and used by plants. Other types of nitrogen-fixing bacteria live in symbiosis in the nodules of the roots of legumes (beans, peas, clover, etc.), supplying the roots with a direct source of ammonia.

Some plants are unable to use ammonia; and instead use **nitrates**. Some bacteria perform **nitrification**, a process which further breaks down ammonia into nitrites (NO_2^-). Then other bacteria convert nitrites into nitrates (NO_3^-).

Nitrogen compounds (such as ammonia and nitrates) are also produced by other physical processes such as volcanic activity. Another source of usable nitrogen is lightning, which reacts with atmospheric nitrogen to form nitrates.

In addition, nitrogen passes along through the food chain, and is recycled through decomposition processes. When plants are consumed, the amino acids are recombined and used, a process that passes the nitrogen-containing molecules on through the food chain or web. Animal waste products, such as urine, release nitrogen compounds (primarily ammonia) back into the environment, yet another source of nitrogen. Finally, large amounts of nitrogen are returned to the Earth by bacteria and fungi, which decompose dead plant and animal matter into ammonia (and other substances), a process known as **ammonification.**

Various species of bacteria and fungi are also responsible for breaking down excess nitrates, a process known as **denitrification**, which releases nitrogen gas back into the air. The nitrogen cycle involves cycling nitrogen through both living and non-living entities.

Carbon Cycle

The carbon cycle is the route by which carbon is obtained, used, and recycled by living things. Carbon is an important element contained in the cells of all species. The study of organic chemistry is the study of carbon-based molecules.

Earth's atmosphere contains large amounts of carbon in the form of carbon dioxide (CO_2). Photosynthetic organisms require the intake of carbon dioxide for the process of photosynthesis, which is the foundation of a food chain. Most of the carbon within organisms is derived from the production of carbohydrates in photosynthetic organisms through photosynthesis. The process of photosynthesis also releases oxygen molecules (O_2), which are necessary for animal respiration. Animal respiration releases carbon dioxide back into the atmosphere in large quantities.

Since plant cells consist of molecules containing carbon, animals that consume photosynthetic organisms are consuming and using carbon from photosynthetic organisms. Carbon is passed along the food chain as these animals are consumed. When animals and photosynthetic organisms die, decomposers, including the detritus feeders (bacteria and fungi) break down the organic matter. Detritus feeders include worms, mites, insects, and crustaceans, which feed on dead organic matter, returning carbon to the cycle through chemical breakdown and respiration.

Carbon dioxide (CO_2) is also dissolved directly into the oceans. It combines with calcium to form calcium carbonate, which is used by mollusks to form their shells. When mollusks die, the shells break down and often form limestone. Limestone is then dissolved by water over time and some carbon may be released back into the atmosphere as CO_2 or used by new ocean species.

Finally, organic matter that is left to decay, may, under certain conditions of heat and pressure, be transformed into coal, oil, or natural gas (the **fossil fuels**). When fossil fuels are burned for energy, the combustion process releases carbon dioxide back into the atmosphere, where it is available to plants for photosynthesis.

Phosphorous Cycle

Phosphorous is another mineral required by living things. Unlike carbon and nitrogen, which cycle through the atmosphere in gaseous form, phosphorous is only found in solid form within rocks and soil. Phosphorous is a key component in ATP, NADP (a molecule that, like ATP, stores energy in its chemical bonds) and many other molecular compounds essential to life.

Phosphorous is found within rocks and is released by the process of erosion. Water dissolves phosphorous from rocks and carries it into rivers and streams. Here phosphorous and oxygen react to form phosphates that end up in bodies of water. Phosphates are absorbed by photosynthetic organisms in and near the water and are used in the synthesis of organic molecules. As in the carbon and nitrogen cycles, phosphorous is then passed through a food chain and returned to the soil through animal waste and organic decay.

New phosphorous enters the cycle as undersea sedimentary rocks are thrust up during the shifting of the Earth's tectonic plates. New rock containing phosphorous is then exposed to erosion and enters the cycling process.

POPULATION GROWTH AND REGULATION

The population growth of a species is regulated by limiting factors that exist within the species' environment. Population growth influences equilibrium in all species under normal conditions because of these limiting factors. A population's overall growth rate is affected by the birth rate (**natality**) and death rate (**mortality**) of the population. The rate of increase within a population is represented by the birth rate minus the death rate. When the birth rate within a population equals the death rate, the population remains at a constant level.

There are two models of population growth, the exponential curve (or J-curve) and the logistic curve (or the S-curve). The exponential curve represents populations in which there is no environmental or social limit on population size, so the rate of growth accelerates over time. Exponential population growth exists only during the initial population growth in a particular ecosystem, since as the population increases, the limiting factors become more influential. In other words, a fish population introduced into a pond would experience exponential population growth until food and space supplies began to limit the population.

The logistic curve reflects the effects of limiting factors on population size, where growth accelerates to a point, then slows down. The logistic curve shows population growth over a longer period of time and represents population growth under normal conditions.

Population growth is directly related to the life characteristics of the population such as the age at which an individual begins to reproduce, the age of death, the rate of growth, etc. For instance, species that grow quickly, mature sexually at an early age, and live a long life would have a population growth rate that exceeds that of species with a short life span and short reproductive span.

In general, most populations have an incredible ability to increase in numbers. A population without limiting factors could overpopulate the world in several generations. It is the limits existing within each ecosystem that keep this from happening.

Limiting Factors

Many factors affect the life of an ecosystem, which may be permanent or temporary. Populations within an ecosystem will be affected by changes in the environment from **abiotic factors** (physical, non-living factors such as fire, pollution, sunlight, soil, light, precipitation, availability of oxygen, water conditions, and temperature) and **biotic factors** (biological factors, including availability of food, competition, predator-prey relationships, symbiosis, and overpopulation).

These biotic and abiotic factors are known as **limiting factors** since they will determine how much a particular population within a community will be able to increase in number. For instance, the resource in shortest supply in an ecosystem may limit population growth. As an example, we know that photosynthetic organisms require phosphorous in order to thrive, so the population growth will be limited by the amount of phosphorous readily available in the environment. Conversely, growth may be limited by having more of a limiting factor (such as heat or water) than it can tolerate. For example, plants need carbon dioxide to grow; however, a large concentration of carbon dioxide in the atmosphere is toxic.

Ecologists now commonly combine these two ideas to provide a more comprehensive understanding of how conditions limit growth of populations. It may be stated that the establishment and survival of a particular organism in an area is dependent upon both 1) the availability of necessary elements in at least the minimum quantity, and 2) the controlled supply of those elements to keep it within the limits of tolerance.

Limiting factors interact with each other and generally produce a situation within the ecosystem that supports homeostasis (a steady-state condition). **Homeostasis** is a dynamic balance achieved within an ecosystem functioning at its optimum level. Homeostasis is the tendency of the ecological community to stay in balance. However, the balance of the ecosystem can be disturbed by the removal, or decrease, of a single factor or by the addition, or increase, of a factor.

Populations are rarely governed by the effect of a single limiting factor; instead, many factors interact to control population size. Changes in limiting factors have a domino effect in an ecosystem; for example, the change in population size of one species will change the dynamics of the entire community. The number of individuals of a particular species living in a particular area is called the population **density** (number of organisms per area).

Both abiotic and biotic limiting factors exist in a single community; however, one may be dominant over the other. Abiotic limiting factors are also known as **density-independent factors**. That is, they are independent of population density. For instance, the populations around Mount St. Helens in Washington state were greatly affected by its eruption in 1980. This effect had nothing to do with the population levels of the area before the eruption. In this situation, the density-independent physical factors dominated the population changes that took place.

Pollution is a major density-independent factor in the health of ecosystems. Pollution is usually a byproduct of human endeavors, and it affects the air or water quality of an ecosystem with secondary effects. In addition to producing pollution, humans may deliberately utilize chemicals such as pesticides or herbicides to limit growth of particular species. These chemicals can damage the homeostatic mechanisms within a community, causing a long-term upset in the balance of an ecosystem.

In other situations, biotic factors, called **density-dependent factors** may be the dominant influence on population in a given area. Density-dependent factors include population growth issues and interactions between species within a community.

Within a given area, there is a maximum level the population may reach at which it will continue to thrive. This is known as the **carrying capacity** of the environment. When an organism has reached the carrying capacity of the ecosystem, the population growth rate will level off and show no net growth. Populations also occupy a particular geographic area with suitable conditions. This total area occupied by a species is known as the **range**. Typically, populations will have the greatest density in the center of their range and lower density at the edges. The area outside the range is known as the area of intolerance for that species, since it is not able to survive there. Environmental changes will affect the size and location of the range, making it a dynamic characteristic.

Over time, species may move in or out of a particular area, a process known as **dispersion**. Dispersion occurs in one of three ways—through **emigration** (permanent one way movement out of the original range), **immigration** (permanent one way movement into a new range), and **migration** (temporary movement out of one range into another, and back). Migration is an important process to many species and communities, since it allows animals that might not survive year-round in a particular ecosystem to temporarily relocate for a portion of the year. Therefore, migration gives the opportunity for greater diversity of species in an ecosystem.

For two or more species living within the same area, their overlap niches (their function in the food chain) are said to be in **competition** if the resource they both require is in limited supply. If the niche overlap is minimal (other sources of food are available), then both species may survive. In some cases, one of the species may be wiped out in an area due to competition, a situation called **competitive exclusion**. This is a rare but plausible occurrence.

A **predator** is simply an organism that eats another. The organism that is eaten is known as the **prey**. The **predator/prey** relationship is one of the most important features of an ecosystem. As seen in our study of the energy cycle, energy is passed from lower trophic levels to higher trophic levels, as one animal is consumed by another. This relationship not only provides transfer of energy up the food chain, it also is a population control factor for the prey species. In situations where natural predators are removed from a region, the overpopulation occurring amongst the prey species can cause problems in the population and community. For instance, the hunting and trapping of wolves in the United States has led to an overpopulation of deer (the prey of wolves) which has caused a shortage of food for deer in some areas causing them to starve.

When two species interact with each other within the same range, it is known as **symbiosis**. **Amensalism** is one type of symbiosis where one species is neither helped nor harmed while it inhibits the growth of another species. **Mutualism** is another form of symbiosis where both species benefit. **Parasitism** is symbiosis in which one species benefits, but the other is harmed. (Parasites are not predators, since the parasitic action takes a long period of time and may not actually kill the host.)

When the entire population of a particular species is eliminated, it is known as **extinction**. Extinction may be a local phenomenon, i.e., the elimination of a population of one species from one area. However, species extinction is a worldwide phenomenon, where all members of all populations of a species die. In fact, the vast majority of species that have ever lived are extinct.

The extinction of a single species may also cause a chain reaction of secondary extinctions if other species depend on the extinct species. Conversely, the introduction of a new species into an area can also have a profound effect on other populations within that area. This new species may compete for the niche of native population or upset a predator/prey balance. For example, the brown tree snake (native to Australia) was introduced into islands in the Pacific years ago. (They probably migrated on ships.) The brown snake has caused the extinction of several species of birds on those Pacific islands. The bird populations could not withstand the introduction of this new predator.

Ultimately, the survival of a particular population is dependent on maintaining a **minimal viable population** size. When a population is significantly diminished in size, it becomes highly susceptible to breeding problems and environmental changes that may result in extinction.

Community Structure

Community structure refers to the characteristics of a specified community, including the types of species that dominate, major climatic trends of the region, and whether the community is open or closed. A **closed community** is one whose populations occupy essentially the same range with very similar distributions of density. These types of communities have sharp boundaries called **ecotones** (such as a pond aquatic ecosystem that ends at the shore). An **open community** has indefinite boundaries, and its populations have varying ranges and densities (such as a forest). In an open community, the species are more widely distributed and animals may actually travel in and out of the area.

An open community is often more able to respond to calamity and may therefore be more resilient. In such a community the boundaries are subtler—the populations of a forest, for instance, may be able to move as necessary to avoid a fire. If, however, a closed community is affected by a traumatic event (for example, a pond being polluted over a short period of time) it may be completely wiped out.

Communities do grow and change over time. Some communities are able to maintain their basic structure with only minor variations for very long periods of time. Others are much more dynamic, changing significantly over time from one type of ecosystem to another. When one community completely replaces another over time in a given area, it is called **succession**. Succession occurs both in terrestrial and aquatic biomes.

Succession may occur because of small changes over time in climate conditions, the immigration of a new species, disease, or other slow-acting factors. It may also occur in direct response to cataclysmic events such as fire, flood, or human intervention (for example, clearing a forest for farmland). The first populations that move back into a disturbed ecosystem tend to be hardy species that can survive in bleak conditions. These are known as **pioneer communities**.

An example of terrestrial succession occurs when a fire wipes out a forest community. The first new colonization will come from quick growing species such as grasses, which will produce over time a grassland ecosystem. The decay of grasses will enrich the soil, providing fertile ground for germination of seeds for shrubs brought in by wind or animals. The shrubs will further prepare the soil for germination of larger species of trees, which over time will take over the shrub-land and produce a forest community once again.

When succession ends in a stable community, the community is known as the **climax community**. The climax community is the one best suited to the climate and soil conditions, and one that achieves a homeostasis. Generally, the climax community will remain in an area until a catastrophic event (fire, flood, etc.) destroys it.

BIOMES

A **biome** is an ecosystem that is generally defined by its climate characteristics. Each biome includes many types of communities interacting within the climatic region. There are several major biomes that have been identified by ecologists. There are two basic types of biome—terrestrial and aquatic.

Terrestrial biomes are those that exist on land; **aquatic biomes** are within large bodies of water. The following table gives the name of the major biomes with their major characteristics:

Biome	Temperature	Precipitation Level	Features
Tropical Rain Forest	warm	high	dense forest, heavy rainfall, abundant vegetation, relatively poor soil
Savanna	warm	moderate	grassland, light seasonal rains
Chaparral	hot summer, temperate winter	low in summer, high in winter	trees, shrubs, small animals, prolonged summer
Temperate Grassland	moderate and seasonal	low for most of year	large land tracts of grassland, shrubs and annuals, rodents, and some larger carnivores
Desert	extreme hot or cold	very low	sandy or rocky terrain, sparse vegetation, mainly succulents, small animals, rodents, reptiles
Tundra	extreme cold	low	modified grassland, permafrost, short growing season with some plants and animals
Taiga	cold	moderate	snow most of year, thick coniferous forests, wide variety of animal life
Temperate Deciduous Forest	moderate, seasonal	moderate	many trees (that lose leaves in cold season), mosses, grasses, shrubs, abundant animal life
Marine Aquatic	varied	not applicable	large amounts of dissolved minerals (particularly salts) in the water, huge array of aquatic animal and plant life
Freshwater Aquatic	varied	not applicable	still or running water with little dissolved minerals, large array of aquatic plant and animal life

Island Biogeography

Biogeography is the study of how photosynthetic organisms and animals are distributed in a particular location; it includes the history of their distribution. Island biogeography is a subdiscipline that investigates the distribution of species in an island habitat. The study of island biogeography is of particular interest to ecologists, since islands are closer to being a closed system (that is, they have less interaction with other ecosystems) than other environments.

Since islands are by nature separated from other land ecosystems, species of both photosynthetic organisms and animals found on a particular island usually have arrived there by natural **dispersal** processes (by air or sea). For instance, there are many plant species with adapted seeds, which will float in water or be carried long distances by air and remain viable. Birds are adapted for dispersion, as are many species of insect, and many sea animals (tortoises, snails, etc.). Dispersal to an island is dependent on geographic as well as historical factors. Distance from other land masses is an important geographic factor. Obviously, the closer the island is to other land, the easier dispersal of species to that island will be. Conversely, islands separated by long stretches of water will have less dispersion. Prevailing winds and ocean currents are also geographic factors that will affect species introduction. Historical factors such as climate shifts (for instance, the shift to an ice age), drought, volcanic action, plate shifting (where the continental plates of the Earth's crust move slowly), etc., will also affect which species are able to travel to a given island. In recent history, dispersion has also occurred through human intervention. Species that inhabit a given ecosystem because humans transported them there are known as **introduced** species.

In some cases, new species develop from parents that were dispersed to the island. These new species are **native** to that island. Arrival of a species on an island, however, does not ensure that it will survive and thrive there. In order to become an established part of the island ecosystem, a species must find a suitable habitat and niche. Ultimately, the species must be able to reproduce for many generations in its new setting, or it will not remain a part of the ecosystem. If the island contains a habitat suitable for the newly arrived species, then that species has a chance of survival in its new environment. Islands may contain several habitats; consequently, islands that support numerous habitats will be more likely to have a wider diversity of species.

Islands may also develop new habitats over time as the climate and geology change, and depending on the size and age of the island. In general, the larger the island and the older the island, the more species it will support. The one exception to this is an old island whose soil has eroded and lost its nutrients. In this case, its habitats may not be able to support life. Also, the harsher the climate (high or low temperature or water conditions) of the island, the fewer species it will have. The geology of the island (whether it is volcanic in origin, for instance) determines the characteristics of the soil, and thus has a direct effect on habitability as well.

PRINCIPLES OF BEHAVIOR

The study of **ethology** involves studying how animals act and react within their environments. Behavior simply is what an organism does and how it does it. Some behavioral characteristics are learned; others are instinctive (inherited).

Behavioral characteristics of animals may include how they acquire food, how they seek out and relate to a mate, how they respond to danger, or how they care for young. Behavior may be as simple as a reflex or may involve responses and interactions between the endocrine, nervous, and musculoskeletal systems.

Some behaviors are extremely simple in nature; they are a response to an environmental stimulus. These basic behaviors are innate; they exist from birth and are genetic in origin (they are inherited). **Innate behaviors** are the actions in animals we call **instincts**. Innate behaviors are highly stereotyped; all individuals of a species perform these behaviors in the same way. **Stereotyped behaviors** are of four basic varieties:

- **taxes** (plural of taxis) are directional responses either toward or away from a stimulus,
- **kineses** are changes in speed of movement in response to stimuli,
- **reflexes** are an automatic movement of a body part in response to a stimulus, and
- **fixed action patterns** (FAP) are complex but stereotyped behaviors in response to a stimulus.

The fixed action pattern is the most complex of stereotyped behaviors. It is a pre-programmed response to a particular stimulus known as a **releaser** or a **sign stimulus**. FAPs include courtship behaviors, circadian rhythms, and feeding of young. Organisms automatically perform FAPs without any prior experience (FAPs are not learned).

Some animal behaviors are learned. **Learned behaviors** may have some basis in genetics, but they also require learning. Generally, there are three types of learned behavior in animals: conditioning, habituation, and imprinting.

Conditioning involves learning to apply an old response to a new stimulus. The classic example of conditioning is that of Pavlov's dogs. Ivan Pavlov was a scientist who studied animal behavior. Dogs have an innate behavior to begin salivating when they see food. Pavlov was able to train dogs to salivate when they heard a bell ring. He trained, or conditioned, them by ringing the bell every time he fed the dogs. The dogs were conditioned to produce an instinctive behavior (salivating) in response to a new stimulus (bell).

B.F. Skinner was another scientist who studied conditioning. Skinner started with the thesis that learning occurs through changes in overt behavior. Skinner believed that when a particular behavior is rewarded, the individual is being conditioned to repeat that behavior. Reinforcement of good behaviors results in repeating that behavior.

Habituation is a learned behavior where the organism produces less and less response as a stimulus is repeated, without a subsequent negative or positive action. For instance, a cat might innately respond to a dog's approach by hissing and raising its hair. However, if the dog regularly approaches, but never attacks, the cat eventually learns that the dog is not a threat and ceases to exhibit the fear behavior. Habituation safeguards species from wasting energy on irrelevant stimuli.

Imprinting is a learned behavior that develops in a critical or sensitive period of the animal's life span. Konrad Lorenz (a behavioral scientist) was able to show that baby geese responded to their mother's physical appearance shortly after birth. During the critical period after birth, the gosling learns to recognize his mother. However, if another object is exposed to the gosling during that critical period (immediately following hatching) the gosling would interpret the substituted object to be its mother. Imprinting generally involves learning a new releaser for an established FAP.

Social Behavior

Some animal species demonstrate **social behavior**—behavior patterns that take into account other individuals. One aspect of animal behavior regards the physical land area where an individual lives. Animals will develop a **home**

range (an area in which they spend most of their time). Animals may also develop an area of land as their **territory**, which lies within the home range, but is the area the individual will defend as his own. The establishment of a territory implies the recognition by one individual that other individuals exist; thus it is a simple social behavior.

Sexual and mating behaviors often rely on complex interactions of the endocrine, nervous, and musculoskeletal systems. In many cases, an individual will compete with another for a particular mate. There are often complex rituals, which are performed before the actual mating experience. These involve many instinctive responses to stimuli and learned behaviors. The sexual and parenting behaviors of animals are social traits that are extremely diverse between species.

In some species, social interactions are highly complex; for instance, an entire population may function as a hierarchy (or society), where individuals have specified roles and status. Insects such as ants and bees, some species of birds, and many primates are among the groups that form societies.

A **society** is an organization of individuals in a population in which tasks are divided, in order for the group to work together. For instance, bees are social insects with a hierarchy (including a queen bee and worker bees). Some individuals are responsible for caring for the queen, others work within the nest, and others gather nectar from outside sources to be brought to the nest. Most of the population is female; the males are called drones and their only responsibility is mating. This division of labor allows the society to perform at a higher function than if each individual acted on its own. Within a society, the individuals may be constantly growing, changing, and adapting, while the functions of the community remain the same over time.

While the social behavior of insects is more a question of division of labor, societies of primates are built around the idea of **dominance**. Older, more established individuals compete for status within the community. A hierarchy is formed through actual competitions among individuals. The community member(s) at the top of the hierarchy enjoy privileges related to their selection of food and mates. This hierarchy is challenged as individuals mature, causing a succession of leaders.

Social behavior is highly dependent on communication within the population and the ability of individuals to adapt behavior according to the needs of the society as a whole. Social animals exhibit a characteristic known as **altruism**, that is, having traits that tend to serve the needs of the society as a whole in addition to its own individual needs.

SOCIAL BIOLOGY

Human Population Growth

As with the study of population growth among other species, human population growth is a direct function of human birth and death rates (natality and mortality). People are able to reason around many of the limiting factors (for example, problems of food shortage or disease) making human population growth a much more complex situation. Furthermore, reproductive behaviors of humans are also subject to the reasoning process, unlike the instinctual mating behaviors of most animals.

The development of vaccines and antibiotics has greatly increased the life span of people in recent history, decreasing the mortality rate. Infant mortality rates have steeply declined in the last 150 years, as safer birthing processes and infant care have been developed. On the other hand, the development of contraceptives has reduced the natality rate in many countries.

Thomas Malthus is one of the most famous human population scientists. In the 1780s Malthus recognized the exponential properties of population growth and calculated that the Earth's food supply would eventually be exhausted by human overpopulation. However, Malthus's calculations did not take into account new technologies allowing for higher yield of food production. So, while the Earth's population has indeed increased exponentially (passing the six billion mark in 2000), and the doubling time (the amount of time it takes for a population to double in size) is decreasing significantly. The Earth has so far been able to support its population for the most part. The Earth has an adequate food supply at present, yet people may currently starve because they are unable to produce food in their region and/or may be politically unable to import food or distribute it properly.

A theory known as **demographic transition** proposes that there are progressive demographic time periods of human population growth. In the first period, birth and death rates are approximately equal, allowing the population to be in equilibrium with the environment. Social evolution (ability to fight disease, mass produce food, etc.) causes the birth rate to overtake the death rate, causing rapid population growth throughout another period. Agrarian lifestyles (where families have numerous children to "work the farm") become less common and children become a liability in urban society. However, **biomedical progress** of urban society causes a lowering of the infant mortality rate. Society then faces a period of dramatic population growth, most of it within cities. The final stage occurs as developed industrialized nations work to lower birth rates through contraceptive practices.

As the human population proceeds through demographic transition, the **age composition** (the relative numbers of individuals of specific ages within the population) changes. As birth rates increase, the population tends to shift toward youth, whereas medical advancements may increase the average age of the population. For instance, in 1900 approximately 40% of Americans were under 18, in 1960, 36%, and in 1996, only 26% of Americans were under 18. Demographic transition also has an effect on the population growth rate.

Meanwhile, the introduction of **genetic engineering** in recent history has produced a complex array of implications. Genetic engineering is the intentional alteration of genetic material of a living organism. Genetic engineering of plant species has produced species able to resist drought, disease, or other threats—providing for more abundant food production. Genetic engineering is also responsible for disease-fighting breakthroughs such as the production of human insulin to fight diabetes. Insulin is produced industrially using genetically engineered bacteria that produce human insulin. The future of genetic engineering, however, is uncertain as we confront the ethical questions regarding the possibility of choosing characteristics of children, or cloning humans.

The growth of human population has also had a profound effect on the biosphere. **Environmental pollution** (the addition of contaminants to the air and water by human intervention and industry) has profoundly affected the ecosystems of the Earth. Most pollution has occurred in recent decades as industrialization has increased.

Progress has been made in the **management of resources** in the recent past. In our discussion of energy and biochemical cycles (early in this chapter), we noted that our biosphere has natural mechanisms that allow for the recycling of energy and nutrients. Careful resource management, including the active human intervention of recycling energy, water, nutrients, and chemicals, will encourage the natural cyclic processes within the biosphere to maintain a viable balance.

CHAPTER 8

Evolution

CHAPTER 8

EVOLUTION

HISTORY OF EVOLUTIONARY CONCEPTS

Evolutionary concepts are the foundation of much of the current study in biology. The term **evolution** refers to the gradual change of characteristics within a population, producing a change in species over time. Evolution is driven by the process of **natural selection**, a feature of population genetics first articulated by Charles Darwin in his book *Origin of Species by Means of Natural Selection, or The Preservation of Favoured Races in the Struggle for Life* (published in 1859). Darwin was the first to explain natural selection as a driving force, and the first to lay out the full range of evidence for evolution. However, scientists before Darwin had already promoted some of the ideas inherent in evolutionary biology.

Carolus Linnaeus, the well-known botanist (who is credited with developing the classification system for organisms still used widely today) speculated on the origin of and relationships between groups of species in the mid-1700s. The French scientist Lamarck proposed that organisms acquire traits over their life span that equip them to survive within their

ORIGIN OF SPECIES.

INTRODUCTION.

WHEN on board H.M.S. 'Beagle,' as naturalist, I was much struck with certain facts in the distribution of the organic beings inhabiting South America, and in the geological relations of the present to the past inhabitants of that continent. These facts, as will be seen in the latter chapters of this volume, seemed to throw some light on the origin of species—that mystery of mysteries, as it has been called by one of our greatest philosophers. On my return home, it occurred to me, in 1837, that something might perhaps be made out on this question by patiently accumulating and reflecting on all sorts of facts which could possibly have any bearing on it. After five years' work I allowed myself to speculate on the subject, and drew up some short notes; these I enlarged in 1844 into a sketch of the conclusions, which then seemed to me probable: from that period to the present day I have steadily pursued the same object. I hope that I may be excused for entering on these personal details, as I give them to show that I have not been hasty in coming to a decision.

My work is now (1859) nearly finished; but as it will take me many more years to complete it, and as my health is far from strong, I have been urged to publish this Abstract. I have more especially been induced to do this, as Mr. Wallace, who is now studying the natural history of the Malay archipelago, has arrived at almost exactly the same general conclusions that I have on the origin of species. In 1858 he sent me a memoir on this subject, with a request that I would forward it to Sir Charles Lyell, who sent it to the Linnean Society, and it is published in the third volume of the Journal of that Society. Sir C. Lyell and Dr. Hooker, who both knew of my work—the latter having read my sketch of 1844—honoured me by thinking it advisable to publish, with Mr. Wallace's excellent memoir, some brief extracts from my manuscripts.

This Abstract, which I now publish, must necessarily be imperfect. I cannot here give references and authorities for my

Introduction to the 1859 edition of Charles Darwin's *Origin of Species*.

environment and pass those traits on to their offspring. He presented the idea that, for example, giraffes developed longer necks during their lifetime from their efforts to reach food high on tree branches. The children, born with longer necks would then further lengthen their necks reaching for high branches, passing these even longer necks on to their children. This Lamarckian theory of acquired characteristics has since been discredited.

Darwinian Concept of Natural Selection

Current theories of evolution have their basis in the work of Charles Darwin and one of his contemporaries, Alfred Russell Wallace. Darwin's book, however, served to catalyze the study of evolution across scientific disciplines. Here is a synopsis of Darwin's ideas:

Population growth and maintenance of a species is dependent on limiting factors. Individuals within the species, that are unable to acquire the minimum requirement of resources, are unable to reproduce. The ecosystem can support only a limited number of organisms—known as the carrying capacity (usually designated by the letter K).

Once the carrying capacity (K) is reached, a competition for resources ensues. Darwin considered this competition to be the basic *struggle for existence*. Some of the competitors will fail to survive. Within every population, there is variation among traits. Darwin proposed that those individuals who win the competition for resources pass those successful traits on to their children. Only the surviving competitors reproduce successfully generation after generation. Therefore, traits providing the competitive edge will be represented most often in succeeding generations.

Modern Concept of Natural Selection

Although the concepts of natural selection put forth by Darwin still form the basis of evolutionary theory today, Darwin had no real knowledge of genetics when he submitted his ideas. Several years after Darwin's writings, Mendel's work (on experimental genetics) was rediscovered independently by three scientists. The laws of genetics served to support the suppositions Darwin had made. Over the next 40 years, the study of genetics included not only individual organisms but also population genetics (how traits are preserved, changed, or introduced within a population of organisms). Progress in the studies of biogeography and paleontology of the early 1900s also served to reinforce Darwin's basic observations.

The modern concept of natural selection emerged from Darwin's original ideas, and is supported today through genetics, population studies, and paleontology. This **modern synthesis** focused on the concept that evolution was a process of gradual (over thousands or hundreds of thousands of generations) adaptive change in traits among populations.

Mechanisms of Evolution

Modern understanding of the process of natural selection recognizes that there are some basic mechanisms that support evolutionary change. Modern theories focus on the change that occurs in entire *gene pools* of species, not among individual populations.

All evolution is dependent upon genetic change. The entire collection of genes within a given population is known as its **gene pool**. Individuals in the population will have only one pair of alleles for a particular single-gene trait. Yet, the gene pool may contain dozens or even hundreds of alleles for this trait. Evolution (meaning a change in allele frequency over time) does not occur through changes from individual to individual, but rather as the gene pool changes through one of a number of possible mechanisms.

One mechanism that drives the changing of traits over time in a species is **differential reproduction**. Natural selection assumes some individuals within a population are more suited for survival, given environmental conditions. Differential reproduction makes this supposition.

The mechanism goes one step further, proposing that those individuals within a population that are most adapted to the environment are also the most likely individuals to reproduce successfully. Therefore, the reproductive processes tend to strengthen the frequency of expression of heritable traits across the population. Differential reproduction increases the number of alleles for desirable traits in the gene pool. This trend will be established and strengthen gradually over time, eventually producing a gene pool in which the heritable trait is more commonly expressed, if environmental conditions remain the same.

Another mechanism of genetic change is mutation. A **mutation** is a change in the DNA sequence of a gene, resulting in a change of the trait. Although a mutation can cause a very swift change in the genotype (genetic code) and possibly phenotype (expressed trait) of the offspring, mutations do not necessarily produce a trait desirable for a particular environment. Mutation is a much more random occurrence than differential reproduction.

Although mutations occur quickly, the change in the gene pool is limited, so change in the population occurs very slowly (over multiple generations). Mutation does provide a vehicle of introducing new genetic possibilities. Genetic traits, which did not exist in the original gene pool, can be introduced through mutation.

Genetic Drift or Neutral Selection

A third mechanism recognized to influence evolution is known as **genetic drift**. Over time, a gene pool (particularly in a small population) may experience a change in frequency of particular genes simply due to chance fluctuations. In a finite population, the gene pool may not reflect the entire number of genetic possibilities of the larger genetic pool of the species. Over time, the genetic pool within this finite population changes, and evolution has occurred. Genetic drift has no particular tie to environmental conditions, and thus the random change in gene frequency is unpredictable. The change of gene frequency may produce a small or a large change, depending on what traits are affected. The process of genetic drift, as opposed to mutation, actually causes a reduction in genetic variety.

Genetic drift occurs within finite separated populations, allowing that population to develop its own distinct gene pool. However, occasionally an individual from an adjacent population of the same species may immigrate and breed with a member of the previously locally isolated group. The introduction of new genes from the immigrant results in a change of the gene pool, known as **gene migration**. Gene migration is also occasionally successful between members of different, but related, species. The resultant hybrids succeed in adding increased variability to the gene pool.

Hardy-Weinberg and Allele Frequencies

The study of genetics shows that in a situation where random mating is occurring within a population (which is in equilibrium with its environment), gene frequencies and genotype ratios will remain constant from generation to generation. This law is known as the **Hardy-Weinberg Law of Equilibrium**, named after the two men (G.H. Hardy and Wilhelm Weinberg, c. 1909) who first studied this principle in mathematical studies of genetics. The Hardy-Weinberg Law is a mathematical formula that shows why recessive genes do not disappear over time from a population.

According to the Hardy-Weinberg Law, the sum of the frequencies of all possible alleles for a particular trait is 1. That is,

$$p + q = 1$$

where the frequency of one allele is represented by **p** and the frequency of another is **q**. It then follows mathematically that the frequency of *genotypes* within a population can be represented by the equation:

$$p2 + 2pq + q2 = 1$$

where the frequency of homozygous dominant genotypes is represented by p2, the homozygous recessive by q2, and the heterozygous genotype by 2pq.

For instance, in humans the ability to taste the chemical phenylthiocarbamide (PTC) is a dominant inherited trait. If **T** represents the allele for tasting PTC and **t** represents the recessive trait (inability to taste PTC), then the possible genotypes in a population would be **TT**, **Tt**, and **tt**. If the frequency of non-tasters (tt) in a particular population is 4% or 0.04 (q2 = 0.04), then the frequency of the allele **t** equals the square root of 0.04, or 0.2. It is then possible to calculate the frequency of the dominant allele, **T**, in the population using the equation:

$$p + 0.2 = 1$$

$$so, p = 0.8$$

The frequency of the allele for tasting PTC is 0.8. The frequency of the various possible genotypes (**TT**, **Tt**, and **tt**) in the population can also be calculated since the frequency of the homozygous dominant genotype (**TT**) is p2 or 0.64 or 64%. The frequency of the heterozygous genotype (**Tt**) is 2pq, and can be calculated once p and q are known.

$$2pq = 2(0.8)(0.2) = 0.32 = 32\%$$

Frequency of TT = 64%, Tt = 32%, tt = 4% . . . totaling 100%
or 0.64 + 0.32 + 0.04 = 1

In order for Hardy-Weinberg equilibrium to occur, the population in question must meet several conditions, random mating (no differential reproduction) must be taking place, very large population size, and no migration, mutation, selection, or genetic drift can be occurring. When these conditions are met, Hardy-Weinberg equilibrium can occur, and there will be no changes in the gene pool over time. Hardy-Weinberg is important to the evolutionary process because it shows that alleles that have no current selective value will be retained in a population over time.

Speciation

A species is an interbreeding population that shares a common gene pool and produces viable offspring. Up to this point we have been considering mechanisms that produce variation within species. It is apparent that to explain evolution on a broad scale, we must understand how genetic change produces new species. There are two mechanisms that produce separate species—allopatric speciation and sympatric speciation.

In order for a new species to develop, substantial genetic changes must occur between populations, which prohibit them from interbreeding. These genetic changes may result from genetic drift or from mutation that take place separately in the two populations. **Allopatric speciation** occurs when two populations are geographically isolated from each other. For instance, a population of squirrels may be geographically separated by a catastrophic event such as a volcanic eruption. Two populations (separated by the volcanic low) continue to reproduce and experience genetic drift and/or mutation over time. This limits each population's gene pool and produces changes in expressed traits. Later, the geographical separation may be eliminated as the volcanic low subsides; even so, the two populations have now experienced too much change to allow them to successfully interbreed again. The result is the production of two separate species.

Speciation may also occur without a geographic separation when a population develops members with a genetic difference, which prevents successful reproduction with the original species. The genetically different members reproduce with each other, producing a population, which is separate from the original species. This process is called **sympatric speciation**.

As populations of an organism in a given area grow, some will move into new geographic areas looking for new resources or to escape predators. In this case, a natural event does not separate the population; instead, part of the population moves. Some of these adventurers will discover new niches and advantageous conditions. Traits that allow this traveling population to use its resources more effectively and to produce more offspring will grow more common over several generations through the process of natural selection. Over time the species will specially adapt to live more effectively in the new environment. Through this process, known as **adaptive radiation**, a single species can develop into several diverse species over time.

All of the above examples describe evolutionary change over a relatively short period of time. While species were traditionally thought to have evolved at a slow, uniform pace over time, scientists have more recently proposed speciation events interspersed within periods of relative stasis. This model is called **punctuated equilibrium**.

Punctuated equilibrium was first proposed as paleontologists studied the fossil record. The older model, **gradualism**, would predict smooth, continuous transitions in the fossil record. However, the fossil record seems to show that organisms in general survive many generations with very little change over long periods of geologic time. New species appeared in the fossils suddenly, without transitional forms, though "sudden" in this context needs to be understood on a geologic time scale. Punctuated equilibrium does not propose "fast" evolutionary change; it would still operate on the scale of tens or hundreds of thousands of years.

Genetic Diversity

As will be noted in the sections to follow, the diversity of species that have evolved on earth is vast. Not only is there great diversity between species, but there are also many distinguishable differences that exist between members of the same species. The genetic variation carried by members of a species for a given trait is known as a **polymorphism**. For example, gender is an example of dimorphism (two variations) in most mammalian species (male and female). Evolutionary mechanisms exist that tend toward a **balanced polymorphism** in order to keep any particular version within a species from dominating, unless that particular version is more suited for its ecological niche (aka being "fitter"). In the example of gender, it is favorable to the species to have an even mix of genders, and the tendency of the gene pool is to favor that balance. If there becomes a shortage of males, the males become more highly sought after and thus "fitter" in the selection process.

Heterozygote advantage is one mechanism whereby polymorphism is maintained. Having two different alleles for a given phenotype is a positive trait for survival in many instances. For example, a well-known situation exists with the gene inheritance of the blood trait for Sickle Cell. The allele HbS is an allele for production of a variant hemoglobin sensitive to oxygen deprival and causing misshapen red blood cells. The allele HbA is the allele for normal hemoglobin production. The HbS allele is found mainly in humans of African and Asian Indian descent. If a person has two HbA alleles, they have normal hemoglobin carrying red blood cells (most of the population). When two HbS alleles are inherited, the person has the disease known as Sickle Cell Anemia, with the accompanying debilitating symptoms and shortened life span. However, a person heterozygous for this trait, HbAHbS, is a carrier for Sickle Cell and has what is known as Sickle Cell Trait. While this trait carries some discomforting symptoms, they are far less debilitating and it also has a positive characteristic; those with Sickle Cell Trait also have a natural resistance to malaria, a deadly illness of the African/Asian regions. For this reason, while the homozygous trait is deadly, the heterozygous trait is advantageous, and therefore is favored, selected, and retained in the population.

A second mechanism of polymorphism is frequency dependent selection which occurs as the frequency of one phenotype increases relative to another.

PLANT AND ANIMAL EVOLUTION

Evolution of the First Cells

The modern theory of the evolution of life on Earth assumes the earliest forms of life began approximately four billion years ago. It is presumed that conditions on Earth were very different from conditions today. The pre-life Earth environment would have been rich in water, ammonia, and methane, all compounds rich in hydrogen. In order for life to arise on Earth, organic molecules, such as amino acids (the building blocks of proteins), sugars, acids, and bases, would need to have formed from the available chemicals.

Over the last century there has been much research into plausible mechanisms for the origin of life. The **Oparin Hypothesis** is one theory regarding the origin of life, developed by a Russian scientist (A.I. Oparin) in 1924. Oparin proposed that the Earth was approximately 4.6 billion years old and that the early Earth had a reducing atmosphere, meaning there was very little free oxygen present. Instead, there was an abundance of ammonia, hydrogen, methane, and steam (H_2O), all escaping from volcanoes. The Earth was in the process of cooling down, so there was a great deal of heat energy available, as well as a pattern of recurring violent lightning storms providing another source of energy. During this cooling of the Earth, much of the steam surrounding the Earth would condense, forming hot seas. In the presence of abundant energy, the synthesis of simple organic molecules from the available chemicals became possible. These organic substances then collected in the hot, turbulent seas (sometimes referred to as the "primordial soup").

As the concentration of organic molecules became very high, they began forming into larger, charged, complex molecules. Oparin called these highly absorptive molecules "coacervates." Coacervates were also able to divide.

Oparin's research involved finding experimental evidence to support his ideas that amino acids combined to form proteins in early Earth conditions. Oparin knew proteins were catalysts so they could have encouraged further change and development of early cells.

Stanley Miller provided support for Oparin's hypotheses in experiments where he exposed simple inorganic molecules to electrical charges similar to lightning. Miller recreated conditions as they were supposed to exist in early Earth history and was successful in his attempt to produce complex organic molecules including amino acids under these conditions. Miller's experiments served to support Oparin's hypotheses.

Sidney Fox, a major evolution researcher of the 1960s, conducted experiments that proved ultraviolet light may have induced the formation of dipeptides from amino acids. Under conditions of moderate dry heat, Fox showed formation of proteinoids, polypeptides of up to 18 amino acids.

He also showed that polyphosphoric acid could increase the yield of these polymers, a process that simulates the modern role of ATP in protein synthesis. These proteins formed small spheres known as microspheres; these showed similarities to living cells.

Further strides were made by researcher Cyril Ponnamperuma who demonstrated that small amounts of guanine formed from the thermal polymerization of amino acids. He also proved the synthesis of adenine and ribose from long-term treatment of reducing atmospheric gases with electrical current.

Once organic compounds had been synthesized, primitive cells most likely developed that contained genetic material in the form of RNA that used energy derived from ATP. These primitive cells were prokaryotic and similar to some bacteria now found on Earth.

The endosymbiont theory suggests that original prokaryotic cells took in other cells that performed various tasks. For instance, an original cell could have absorbed several symbiotic bacteria that then evolved into mitochondria. Several cells that lived in symbiosis would have combined and evolved to form a single eukaryotic cell.

Plant Evolution

The evolution of plant species is considered to have begun with heterotrophic prokaryotic cells. Since it is presumed that the early Earth's atmosphere was lacking in oxygen, early cells were anaerobic. Over time, some bacteria evolved the ability to carry on photosynthesis (cyanobacteria), thus becoming autotrophic, which in turn introduced significant amounts of oxygen into the atmosphere. As oxygen is poisonous to most anaerobic cells, a new niche opened up: cells able not only to survive in the presence of oxygen, but also to use it in metabolism.

Cyanobacteria were incorporated into larger aerobic cells, which then evolved into photosynthetic eukaryotic cells. Cellular organization increased, nuclei and membranes formed, and cell specialization occurred, leading to multicellular photosynthetic organisms, that is, plants.

The earliest plants were aquatic, but as niches filled in marine and freshwater environments, plants began to move onto land. Anatomical changes occurred over time allowing plants to survive in a nonaqueous environment. Cell walls thickened, and tissues to carry water and nutrients developed. As plants continued to adapt to land conditions, differentiation of tissues continued, resulting in the evolution of stems, leaves, roots, and seeds. The development of the seed was a key factor in the survival of land plants. Asexual reproduction dominated in early species, but sexual reproduction developed over time, increasing the possibilities of diversity.

Processes of adaptive radiation, genetic drift, and natural selection continued over long periods of time to produce the incredible diversity seen in the plant world today.

Animal Evolution

The evolution of animals is thought to have begun with marine protists. Although there is no fossil record, going back to the protist level, animal cells bear the most similarity to marine protist cells. Fossilized burrows from multicellular organisms begin to appear in the geological record approximately 700 million years ago, during the Precambrian period. These multicellular animals had only soft parts—no hard parts, which could be fossilized.

During the Cambrian period (the first period of the Paleozoic Era), beginning about 570 million years ago, the fossil record begins to show multicellular organisms with hard parts, namely exoskeletons. The fossil record at this time includes fossil representations from all modern-day (and some extinct) phyla. This sudden appearance of multitudes of differentiated animal forms is known as the **Cambrian explosion**.

At the end of the Paleozoic Era, the fossil record attests to several mass extinction events. These combined events resulted in the extinction of about 95% of animal species developed to this point. Fossils indicate that many organisms, such as Trilobites, that were numerous in the Cambrian era, did not survive the end of the Paleozoic Era.

Approximately 505 million years ago marked the beginning of the Ordovician period, which lasted until about 440 million years ago. The Ordovician period was marked by diversification among species that survived past the Cambrian extinctions. The Ordovician is also known for the development of land plants. Early forms of fish arose in the Cambrian, but developed during the Ordovician, these being the first vertebrates to be seen in the fossil record. Again, the end of the Ordovician is marked by vast extinctions, but these extinctions allowed the opening of ecological situations, which in turn encouraged adaptive radiation.

Adaptive radiation is the mechanism credited with the development of new species in the next period, the Silurian, from 440 to 410 million years ago. The Silurian period is marked by widespread colonization of landmasses by plants and animals. Large numbers of insect fossils are recognizable in Silurian geologic sediments, as well as fish and early amphibians. The mass movement onto land by formerly marine animals required adaptation in numerous areas, including gas exchange, support (skeletal), water conservation, circulatory systems, and reproduction.

In the study of animal evolution, attention is paid to two concepts: homology and analogy. Structures that exist in two different species because they share a common ancestry are called **homologous**. For instance, the forelimbs of a salamander and an opossum are similar in structure because of common ancestry. **Analogous** structures are similar because of their common function, although they do not share a common ancestry. Analogous structures are the product of **convergent evolution**. For instance, birds and insects both have wings. Although they are not relatives, the wings evolved as a result of convergence. Convergence occurs when a particular characteristic evolves in two unrelated populations. Wings of insects and birds are analogous structures (they are similar in function regardless of the lack of common ancestors).

The process of **extinction** has played a large part in the direction evolution has taken. Extinctions occur at a generally low rate at all times. It is presumed that species that face extinction have not been able to adapt appropriately to environmental changes. However, there have also been several "extinction events" that have wiped out up to 95% of the species of their time. These events served to open up massive ecological niches, encouraging evolution of multitudes of new species.

Approximately 400 million years ago, the first amphibians gave rise to early reptiles that then diversified into birds, then mammals. One branch of mammals developed into the tree-dwelling primates, considered the ancestors of humans.

Human Evolution

Humans are thought to have evolved from primates who, over time, developed larger brains. A branch of bipedal primates gave rise to the first true hominids about 4.5 million years ago. The earliest known hominid fossils were found in Africa in the 1970s. The well-known "Lucy" skeleton was named *Australopithecus afarensis*. It was determined from the skeleton of *Australopithecus* that it was a biped. It had a human-like jaw and teeth, but a skull that was more like that of a small ape. The arms were proportionately longer than humans', indicating the ability to still be motile in trees.

The fossilized skulls of *Homo erectus*, the oldest known fossil of the human genus, is thought to be about 1.8 million years old. The skull of *Homo erectus* was quite a lot larger than *Australopithecus*. It was approximately the size of a modern human brain. *Homo erectus* was thought to walk upright and had facial features more closely resembling humans than apes. The oldest fossils to be designated *Homo sapiens* are also called Cro-Magnon man, with brain size and facial features essentially the same as modern humans. Cro-Magnon *Homo sapiens* are thought to have evolved in Africa and migrated to Europe and Asia approximately 100,000 years ago.

Female Cro-Magnon Skull

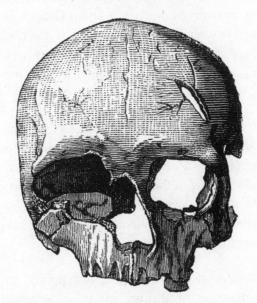

Source: Wikipedia

Evolutionary Ecology

Organisms evolve within ecosystems; therefore, the ecological circumstances affect (if not determine) the course of evolution of species in a particular area. Some organisms are better suited to develop in a new ecosystem. Others only thrive in an established equilibrium. The characteristics that differentiate these types of organisms are known collectively as **life history strategies**. There are two types of life history strategies: opportunistic and equilibreal.

Organisms with **opportunistic** life history strategies (also known as r-selected) tend to be pioneer species in a new or recently devastated community. In addition to traits that allow them to succeed in the long term, they also have traits that help make them succeed in a changing or new ecosystem. They tend to have short maturation times and short overall life spans. They tend to have high mortality rates. Often reproduction is asexual, with high numbers of offspring. They find it easy to disperse over large areas. They do not parent their young. These are rapidly reproducing species that are also easily wiped out by more sophisticated populations that follow. For example, dandelions are an opportunistic species.

Species with **equilibreal** life strategies (also known as K-selected) are those organisms that overtake the opportunistic pioneer species. These tend to have long life spans with a long maturation time and corresponding low mortality rate. They reproduce sexually and produce fewer (longer living) offspring that they tend to parent. They tend to stay within their established borders rather than dispersing. These characteristics form the basis for particular species to dominate in varying ecosystems. For example, an oak tree is an equilibreal species.

One of the most interesting ecological behaviors to explain through evolution is **altruism** (social behavior in which organisms seem to place the needs of the community over their own need). An altruistic trait may actually decrease the fitness of the individual with the trait (known as the cost of altruism), while it increases the fitness of the community (the benefit of altruism). When you look at altruism in terms of the individual, it would not have been an evolved trait since it decreases the individual's fitness.

However, when looked at in terms of the community, an altruistic trait has value. In order for the traits of altruism to evolve, it would be necessary for some other factor to influence the preservation and proliferation of those traits. This is thought to occur in nature through a process known as **kin selection**.

Kin selection is the tendency of an individual to be altruistic toward a close relative, resulting in the preservation of its genetic traits. Close relatives have a greater likelihood of passing on identical traits to their offspring. For instance, kin selection for a gene that causes an animal to share food with its close relatives would result in this altruistic trait being spread throughout the gene pool and passed on to future generations. Thus, those relatives are more likely to survive, and their genes are passed on to the offspring, preserving the altruistic trait in future generations.

Since the communities that have altruistic individuals are more likely to persevere, natural selection will work to maintain those communities, while the weaker communities die out. It is widely accepted that those communities containing altruistic individuals are made up of close relatives that have been able to preserve the altruism through kin selection.

Diversity of Life

The world as we know it involves an array of organisms that some would consider immeasurable. In the field of biology, however, we seek to identify and understand our world and its inhabitants. It is out of this desire for knowledge that biologists from the past developed methods for categorizing and classifying all of the diverse living things that they found to exist.

Classification of Living Organisms

The study of **taxonomy** seeks to organize living things into groups based on morphology, or more recently, genetics. Scientists have sought to categorize the great diversity of life on Earth for hundreds of years.

Carolus Linnaeus, who published his book *Systema Naturae* in 1735, first developed our current methods of taxonomy. Linnaeus based his taxonomic keys on the **morphological** (outward anatomical) differences seen among species. Linnaeus designed a system of classification for all known and unknown organisms according to their anatomical similarities and differences. Although Linnaeus was a Biblical creationist who sought to show the great diversity of creation, and although he devised his system over 100 years before Darwin's *Origin of Species*, his system remains as the basis of our classification system today.

Linnaeus used two Latin-based categories—*genus* and *species*—to name each organism. Every genus name could include one or more types of species. We refer to this two-word naming of species as **binomial nomenclature** (literally meaning "two names" in Latin). For example, Linnaeus named humans *Homo sapiens* (literally "man who is wise"). *Homo* is the genus name and *sapiens* the species name. *Homo sapiens* is the only extant species left from the genus *Homo*.

Beyond genus and species, Linnaeus further categorized organisms in a total of seven levels. Every **species** also belongs to a **genus**, **family**, **order**, **class**, **phylum**, and **kingdom**. *Kingdom* is the most general category, *species* the most limited. Taxonomists now also add "sub" and "super" categories to give even more opportunity for grouping similar organisms, and have added categories even more general than kingdom (called **domains**).

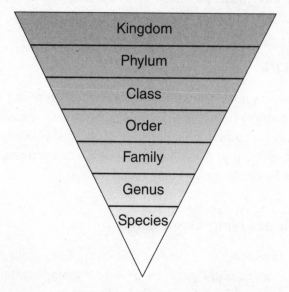

The most modern classification system contains three domains: the **Archaea,** the **Eubacteria**, and the **Eukaryota**. The organisms of the domain Archaea are prokaryotic, have unique RNA, and are able to live in the extreme ecosystems on Earth. The domain Archaea includes methane-producing organisms, and organisms able to withstand extreme temperatures and high salinity. The domain Eubacteria contains the prokaryotic organisms we call bacteria.

[Note: Some taxonomists still use a system including five (or six) kingdoms and no domains. In this case, the kingdom Monera would include organisms that are considered by other taxonomists to be included within the domains Archaea and Eubacteria.]

Kingdoms of the Domain Eukaryota

Kingdom	# Known Phyla/Species	Nutrition	Structure	Included Organisms
Protista	27/250,000 +	photosynthesis, some ingestion and absorption	large eukaryotic cells	algae & protozoa
Fungi	5/100,000 +	absorption	multicellular (eukaryotic) filaments	mold, mushrooms, yeast, smuts, mildew
Animalia	33/1,000,000 +	ingestion	multicellular specialized eukaryotic motile cells	various worms, sponges, fish, insects, reptiles, amphibians, birds, and mammals
Plantae	10/250,000 +	photosynthesis	multicellular specialized eukaryotic non-motile cells	ferns, mosses, woody and non-woody flowering plants

The preceding chart gives the major features of the four kingdoms of the Eukaryota. The domain Eukaryota includes all organisms that possess eukaryotic cells. The domain Eukaryota includes the four kingdoms: **Kingdom Protista, Kingdom Fungi, Kingdom Animalia,** and **Kingdom Plantae**.

There are nine major phyla within the **Kingdom Animalia**. The Phyla are as follows:

1. **Porifera**—the sponges
2. **Cnidaria**—jellyfish, sea anemones, hydra, etc.
3. **Platyhelminthes**—flat worms
4. **Nematoda**—round worms
5. **Mollusca**—snails, clams, squid, etc.
6. **Annelida**—segmented worms (earthworms, leeches, etc.)
7. **Arthropoda**—crabs, spiders, lobster, millipedes, insects

8. **Echinodermata**—sea stars, sand dollars, etc.

9. **Chordata**—fish, amphibians, reptiles, birds, mammals, lampreys

Vertebrates are within the phylum Chordata, which is split into three subphyla: the **Urochordata** (animals with a tail cord such as tunicates), the **Cephalochordata** (animals with a head cord, such as lampreys), and **Vertebrata** (animals with a backbone).

The subphylum Vertebrata is divided into two **superclasses**: the **Aganatha** (animals with no jaws) and the **Gnathostomata** (animals with jaws). The Gnathostomata includes six classes with the following major characteristics:

a. **Chondrichthyes**—fish with a cartilaginous endoskeleton, two-chambered heart, 5–7 gill pairs, no swim bladder or lung, and internal fertilization (sharks, rays, etc.).

b. **Osteichthyes**—fish with a bony skeleton, numerous vertebrae, swim bladder (usually), two-chambered heart, gills with bony gill arches, and external fertilization (herring, carp, tuna).

c. **Amphibia**—animals with a bony skeleton, usually with four limbs having webbed feet with four toes, cold-blooded (ectothermic), large mouth with small teeth, three-chambered heart, separate sexes, internal or external fertilization, amniotic egg (salamanders, frogs, etc.).

d. **Reptilia**—horny epidermal scales, usually have paired limbs with five toes (except limbless snakes), bony skeleton, lungs, no gills, most have three-chambered hearts, cold-blooded (ecothermic), internal fertilization, separate sexes, mostly egg-laying (oviparous), eggs contain extra-embryonic membranes (snakes, lizards, alligators).

e. **Aves**—spindle-shaped body (with head, neck, trunk, and tail), long neck, paired limbs, most have wings for flying, four-toed feet, feathers, leg scales, bony skeletons, bones with air cavities, beaks, no teeth, four-chambered hearts, warm blooded (endothermic), lungs with thin air sacs, separate sexes, egg-laying. Eggs have hard calcified shell (birds, ducks, sparrows, etc.).

f. **Mammalia**—body covered with hair, glands (sweat, scent, sebaceous, mammary), teeth, fleshy external ears, usually four limbs, four-chambered heart, lungs, larynx, highly developed brain, warm-blooded, internal fertilization, live birth (except for the egg-laying monotremes), milk producing (cows, humans, platypus, apes, etc.).

PRACTICE TEST 1

CLEP Biology

Also available at the REA Study Center (*www.rea.com/studycenter*)

This practice test is also offered online at the REA Study Center. All CLEP exams are computer-based, and our test is formatted to simulate test-day conditions. We recommend that you take the online version of the test to receive these added benefits:

- **Timed testing conditions** – helps you gauge how much time you can spend on each question
- **Automatic scoring** – find out how you did on the test, instantly
- **On-screen detailed explanations of answers** – gives you the correct answer and explains why the other answer choices are wrong
- **Diagnostic score reports** – pinpoint where you're strongest and where you need to focus your study

PRACTICE TEST 1

CLEP Biology

(Answer sheets appear in the back of the book.)

TIME: 90 Minutes
115 Questions

DIRECTIONS: Each of the questions or incomplete statements below is followed by five possible answers or completions. Select the BEST choice in each case and fill in the corresponding oval on the answer sheet.

1. A small non-protein substance such as iron that works with enzymes to promote catalysis is known as

   (A) a mineral
   (B) an inorganic cofactor
   (C) a coenzyme
   (D) a hormone
   (E) an ion

2. A particular plant has individuals that are either male or female. A male individual of this plant may have all of the following EXCEPT

   (A) a filament
   (B) an anther
   (C) a stigma
   (D) pollen grains
   (E) tube nuclei

3. The cells of which of the following organisms are prokaryotic?

   (A) Mold
   (B) Seaweed
   (C) Blue-green bacteria
   (D) Fern
   (E) Hydra

4. Which of the following represents a plausible progression in the evolution of plants?

(A) autotrophic eukaryotic cells $\Rightarrow$ aerobic prokaryotic cells $\Rightarrow$ photosynthetic cells $\Rightarrow$ multicellular plants
(B) heterotrophic eukaryotic cells $\Rightarrow$ anaerobic prokaryotic cells $\Rightarrow$ autotrophic cyanobacteria $\Rightarrow$ multicellular plants
(C) aerobic eukaryotic cells $\Rightarrow$ anaerobic eukaryotic cells $\Rightarrow$ photosynthetic cells $\Rightarrow$ multicellular plants
(D) anaerobic prokaryotic cells $\Rightarrow$ aerobic prokaryotic cells $\Rightarrow$ anaerobic eukaryotic cells $\Rightarrow$ multicellular plants
(E) anaerobic prokaryotic cells $\Rightarrow$ autotrophic cyanobacteria $\Rightarrow$ aerobic eukaryotic cells $\Rightarrow$ multicellular plants

5. Legumes perform a unique ecological function by

(A) hosting nitrogen-fixing bacteria in their root nodules
(B) providing food for primary consumers
(C) releasing oxygen into the atmosphere
(D) releasing carbon dioxide into the atmosphere
(E) attracting lightning

6. Which of the following is found in a carbohydrate molecule?

(A) N
(B) S
(C) NH_2
(D) PO_4
(E) CH_2O

7. Each of the following statements regarding the hydrologic cycle is true EXCEPT:

(A) Water evaporates from bodies of water and plant surfaces and forms clouds in the atmosphere.
(B) The hydrologic cycle does not affect any of the other biogeochemical cycles.
(C) Water is released from clouds as precipitation.
(D) Water vapor in the atmosphere protects the Earth from rapid and extreme temperature changes.
(E) Runoff occurs, directing water into waterways that eventually empty into the oceans.

8. Which of the following is a monosaccharide?

   (A) Glucose
   (B) Cellulose
   (C) Table sugar
   (D) Amylase
   (E) Starch

Questions 9–11 consist of five lettered terms followed by a list of numbered phrases. For each numbered phrase, select the one term that is most closely related to it. Each term may be used once, more than once, or not at all.

   (A) Releaser
   (B) Instinct
   (C) Stimulus
   (D) Reflex
   (E) Fixed action pattern

9. Highly stereotyped innate behavior

10. Automatic movement in response to an environmental signal

11. Unlearned series of actions that are a pre-programmed but complex response to a particular environmental signal

12. The study of the interaction of organisms with their living space is known as

   (A) environmentalism
   (B) habitology
   (C) zoology
   (D) ecology
   (E) paleontology

13. Which of the following is likely to happen when a limited amount of enzyme is added to a reaction with an unlimited amount of substrate?

   (A) The rate of the reaction increases, then levels off as all of the enzyme is engaged.
   (B) The rate of the reaction rises steeply, then the reaction stops completely.
   (C) The reaction does not occur.
   (D) The rate of the reaction rises steeply and continues to rise.
   (E) The rate of the reaction slowly decreases.

14. Which of the following organs does not function in immunity to defend the body from infection?

    (A) Tonsils
    (B) Lymph nodes
    (C) Spinal cord
    (D) Spleen
    (E) Thymus

15. All of the statements about the following reaction are true EXCEPT:

glycerol + 3 fatty acids = fat (triglyceride) + 3H₂O

    (A) It is an exothermic reaction.
    (B) It is a combination reaction.
    (C) It is an endothermic reaction.
    (D) This reaction occurs within animals in order to store energy.
    (E) Water is a by-product.

16. A cell without a nucleus or membrane-bound organelles is

    (A) prokaryotic
    (B) replicating
    (C) in telophase
    (D) eukaryotic
    (E) endocytic

17. All of the following are found in the cells of fungi EXCEPT

    (A) chloroplasts
    (B) nucleus
    (C) mitochondria
    (D) plasma membranes
    (E) ribosomes

18. Grana are embedded within which part of the chloroplast?

    (A) Stroma
    (B) Pigments
    (C) Chlorophyll
    (D) Carotene
    (E) Christea

19. The ultimate source of energy for most life on Earth is

    (A) water
    (B) protein
    (C) consumers
    (D) the Sun
    (E) ATP

20. The conversion of light energy into chemical energy is accomplished by

    (A) catabolism
    (B) oxidative phosphorylation
    (C) metabolism
    (D) protein synthesis
    (E) photosynthesis

21. The sum total of a species' genetic information is known as its

    (A) genes
    (B) chromosomes
    (C) polyploidy
    (D) inheritance
    (E) genome

22. All of the following are steps in the translation part of protein synthesis EXCEPT:

    (A) Free bases line up along the DNA template and are bonded together forming a single strand of RNA.
    (B) A ribosome attaches to start codon on mRNA and links a tRNA with its attached amino acid.
    (C) The ribosome continues to link a sequence of tRNA molecules that correspond with the mRNA strand being encoded.
    (D) The amino acids are linked by ribosomal enzymes into a protein chain.
    (E) A terminating codon stops the synthesis process and releases the newly formed protein.

23. Transduction refers to the process whereby a bacterium's genetic makeup is altered when

    (A) genetic material is absorbed from dead cell debris in the cell's environment
    (B) some genetic material is transferred from one bacterium to another via a viral bacteriophage
    (C) genetic expression is suppressed by regulatory genes
    (D) mutations occur within genetic material
    (E) a bacterium's chromatin is destroyed by high temperatures

24. Chromosomes that are paired with others of similar size and shape within the nucleus are known as

    (A) homologs
    (B) histones
    (C) dialogs
    (D) genes
    (E) nucleosomes

25. Restriction enzymes cut samples of DNA into fragments by

    (A) binding to specific sequences of nucleotides and breaking the sugar-phosphate backbone
    (B) unwinding the DNA
    (C) breaking the base-to-base hydrogen bonds
    (D) transcribing the RNA sequences
    (E) oxidizing the DNA strands

Questions 26–30 consist of five lettered terms followed by a list of numbered phrases. For each numbered phrase select the one term that is most closely related to it. Each term may be used once, more than once, or not at all.

(A) Prophase
(B) Metaphase
(C) Anaphase
(D) Telophase
(E) Cytokinesis

26. Nuclear membrane forms around new groups of single-stranded chromosomes

27. Cytoplasm splits forming two distinct cells

28. Chromatin condenses, centrioles move to opposite ends of the cell, nuclear membrane dissolves, kinetochore forms

29. Chromosomes align along the equatorial plane of the cell

30. Paired chromosomes separate at the kinetochore, each chromosome travels along the spindle fibers to opposite ends of the cell

31. Each of the following statements about meiosis is true EXCEPT:

(A) Meiosis produces two exact replica daughter cells.
(B) Meiosis occurs in reproductive organs to produce gametes.
(C) The first phase of meiosis is known as reduction, which reduces the ploidy from 2n to n (diploid to haploid).
(D) The second phase of meiosis division produces four haploid daughter cells, each with a different combination of chromosomes.
(E) Meiosis begins with chromosome duplication.

32. The iron-containing molecule that carries oxygen within red blood cells throughout the body via the circulatory system is

(A) lymphocyte
(B) erythrocyte
(C) eosinophil
(D) hemoglobin
(E) neutrophils

33. When sodium (Na$^+$) ion concentration outside a cell increases, water molecules travel out of the cell through the cell membrane. This process is known as

    (A) osmosis
    (B) facilitated diffusion
    (C) active transport
    (D) exocytosis
    (E) endocytosis

34. The _____ is the organ that prevents food from entering the bronchial tubes.

    (A) glottis
    (B) epiglottis
    (C) trachea
    (D) larynx
    (E) pharynx

35. The site of transfer for nutrients, water, and wastes between a mammalian mother and embryo is the

    (A) yolk sac membrane
    (B) uterus
    (C) placenta
    (D) allantois
    (E) gastrula

36. The function of the gall bladder and pancreas is to aid digestion by producing digestive enzymes and secreting them into

    (A) the small intestine
    (B) the large intestine
    (C) the stomach
    (D) the esophagus
    (E) the colon

37. Electrical shock can restart a heart that has stopped beating. Which of the following statements is a valid reason for this fact?

   (A) Electric shock stimulates the nervous system.
   (B) Electric shock causes smooth muscle to contract.
   (C) Electric shock pushes blood through a stopped heart.
   (D) Electric shock stimulates cardiac muscle causing it to contract.
   (E) Electric shock forces air into lungs.

38. All of the following may cause mutations in a DNA sequence EXCEPT

   (A) X-rays
   (B) chemical exposure
   (C) random copying error
   (D) sunlight
   (E) crossing over

Questions 39–43 consist of five lettered terms followed by a list of numbered phrases. For each numbered phrase, select the one term that is most closely related to it. Each term may be used once, more than once, or not at all.

   (A) Primary oocytes
   (B) Egg cell
   (C) Secondary spermatocytes
   (D) Zygote
   (E) Polar body

39. Present in reproductive organs at birth

40. Female haploid cell that is ready for fertilization

41. Haploid cells that will develop into male gametes

42. Infertile cell resulting from meiosis II in females

43. Undergo meiosis II to form spermatid

44. Bones perform many functions in the human body. All of the following are functions of bones EXCEPT they

   (A) provide structure and support
   (B) provide protection for organs
   (C) produce red blood cells
   (D) aid in locomotion
   (E) produce calcium and phosphate

Questions 45–49 In the illustration below, "T" stands for tall and "t" for short.

	T	T
t	Tt	Tt
t	Tt	Tt

45. The illustration above is called

   (A) a Mendelian diagram
   (B) a genotype
   (C) a phenotype
   (D) a phenogram
   (E) a Punnett square

46. What is the genotype of each of the parents?

   (A) One parent has the genotype TT and the other tt.
   (B) Both have the genotype Tt.
   (C) Both have the genotype TT.
   (D) Both have the genotype tt.
   (E) One parent has the genotype Tt and the other TT.

47. The gametes for this cross will have which possible alleles?

   (A) T, T
   (B) T, t
   (C) TT, tt
   (D) t, t
   (E) TtTt, TtTt

48. What will be the phenotypic ratio of the offspring in this cross?

   (A) 4 tall: 0 short
   (B) 2 tall: 2 short
   (C) 4 Tt: 0 tt
   (D) 3 Tt: 1 tt
   (E) 0 Tt: 4tt

49. If two of the offspring are crossed, what will the phenotypic ratio of the next generation be?

   (A) 4 Tt: 0 tt
   (B) 1 tt: 2 Tt; 1 TT
   (C) 4 tall: 0 short
   (D) 3 tall: 1 short
   (E) 0 tall: 4 short

50. All of the following are major structural regions of plant roots EXCEPT

   (A) the meristematic region
   (B) the elongation region
   (C) the root cap
   (D) the epistematic region
   (E) the maturation region

51. Water molecules are attracted to each other due to which of the following?

   (A) Polarity, partial positive charge near hydrogen atoms, partial negative charge near oxygen atoms
   (B) Inert properties of hydrogen and oxygen
   (C) Ionic bonds between hydrogen and oxygen
   (D) The crystal structure of ice
   (E) Brownian motion of hydrogen and oxygen atoms

52. The Law of Segregation in Genetics states that

   (A) one gene is usually dominant over the other (expresses itself over the other)
   (B) genes are inherited via a process of multiple alleles
   (C) genes are randomly separated in gamete formation and brought together in fertilization
   (D) gamete formation causes mutation in genetic material of both parents
   (E) offspring always have the same genotype as one of the parents

53. All of the following are true EXCEPT:

    (A) Animal cells have organized nuclei and membrane-bound organelles.
    (B) Animal cells do not have cell walls or plastids.
    (C) Animals only reproduce asexually.
    (D) Animals develop from embryonic stages.
    (E) Animals are heterotrophic (they do not produce their own food).

54. All of the following terms represent the movement of a species in or out of a given area EXCEPT

    (A) competition
    (B) emigration
    (C) dispersion
    (D) immigration
    (E) migration

55. Sharks and dolphins have similar body shapes. Which of the following is the most likely explanation of this fact?

    (A) Sharks and dolphins have a common ancestry.
    (B) The similar body shape traits are the result of convergent evolution.
    (C) Sharks and dolphins are competitors in their ocean communities.
    (D) A species of shark evolved into the first dolphin species.
    (E) The similar body shape traits are the result of allopatric speciation.

Questions 56–60: Match the tissue type (all are types of primary root tissue) labeled a–e (and corresponding to answer choices A–E) on the diagram of a root cross-section below with the correct name given in the following five numbered questions.

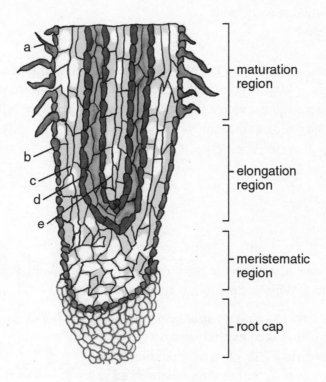

Root Cross-section

(A) a

(B) b

(C) c

(D) d

(E) e

56. Vascular cylinder

57. Epidermis

58. Root hairs

59. Endodermis

60. Cortex

61. The term *Homo sapiens* is an example of

    (A) family and phylum names
    (B) a class name
    (C) a taxonomic key
    (D) binomial nomenclature
    (E) species

62. In humans, the ability to roll the tongue is an inherited trait and the allele for tongue-rolling is dominant. If 36% of the population cannot roll their tongue, what is the frequency of the heterozygous genotype within the population, according to the Hardy-Weinberg Equation?

    (A) 0.06
    (B) 0.04
    (C) 0.48
    (D) 0.52
    (E) 0.36

63. An unknown plant found in the forest has five petals on its flower and a tap root system. Which of the following is most likely TRUE?

    (A) The stem has random arrangements of vascular bundles.
    (B) The plant's seed has only one cotyledon.
    (C) The leaves of the plant have parallel veins.
    (D) The leaves of the plant have networked veins.
    (E) The plant is male.

64. Members of which of the following categories are most closely related?

    (A) Phylum
    (B) Genus
    (C) Kingdom
    (D) Class
    (E) Order

65. Which of the following is NOT included in the Oparin Hypothesis?

(A) H_2O once existed only in the form of ice.
(B) The young Earth had very little oxygen present in the atmosphere.
(C) The Earth is more than 4 billion years old.
(D) Heat energy was abundantly available because of the Earth's cooling.
(E) Large organic molecules capable of dividing and absorption (coacervates) became plentiful, resulting in eventual evolution of early life.

66. Which of the following is an explanation of how altruistic traits evolve?

(A) The presence of an altruistic trait in an individual decreases its fitness to survive.
(B) The process of kin selection preserves altruistic traits.
(C) Genetic drift accounts for the development of altruistic traits.
(D) Species that experience adaptive radiation also develop altruism.
(E) Altruistic traits are always dominant.

67. *Streptococcus pyogenes* bacteria cause throat infections in humans, but can be killed with the antibiotic penicillin. If penicillin therapy is not administered correctly, some bacteria may survive. The surviving bacteria are those with a higher level of resistance to penicillin. The living resistant bacteria will reproduce, magnifying the traits of resistance in subsequent generations. This is an example of

(A) genetic drift
(B) natural selection
(C) mutation
(D) genetic equilibrium
(E) allopatric speciation

68. The major driving force of the evolution of species is known as

(A) the Oparin Hypothesis
(B) natural selection
(C) allopatric speciation
(D) the Hardy-Weinberg Equilibrium
(E) genetic drift

Questions 69–73 consist of five lettered terms followed by a list of numbered phrases. For each numbered phrase, select the one term that is most closely related to it. Each term may be used once, more than once, or not at all.

(A) Adaptive radiation
(B) Sympatric speciation
(C) Punctuated equilibrium
(D) Kin selection
(E) Genetic drift

69. Short period of quick mutation and change resulting in new species

70. Change in frequency of particular genes in a population over time due to chance fluctuations

71. Process whereby one species can evolve into several new species over time as migration to new areas occurs and traits are specialized to fit new habitat

72. The development of members within a population that possess differences preventing successful reproduction with the original population

73. The tendency for an individual to express altruistic traits toward close relatives, thus preserving the genes that produce altruistic traits

74. The biogeography of a tropical island is affected by all of the following factors EXCEPT

(A) volcanic activity
(B) distance from other landmasses
(C) human population
(D) prevailing winds
(E) fossil preservation within the geologic column

75. A form of symbiosis in which one species benefits while the other is harmed is called

(A) parasitism
(B) mutualism
(C) amensalism
(D) predation
(E) habituation

76. All of the following are steps in the carbon cycle EXCEPT:

(A) Carbon is taken in by plants and used to form carbohydrates through photosynthesis.
(B) Carbon dioxide is dissolved out of the air into ocean water and combined with calcium to form calcium carbonate.
(C) Carbon dioxide is taken in by animal respiration and used to form carbohydrates.
(D) Detritus feeders return carbon contained in organic compounds to elemental form.
(E) Burning fossil fuels release carbon dioxide into the atmosphere where it can be used by plants.

77. Large protein molecules may be secreted from a cell by the process of

(A) endocytosis
(B) exocytosis
(C) diffusion
(D) active transport
(E) facilitated diffusion

78. The enzyme amylase is present in saliva and is instrumental in the breakdown of starches in early digestion. Which of the following is the most likely reason for amylase's suitability to aid in the catalysis of starches?

(A) The shape of the active site on the amylase molecule matches the shape of starch molecules.
(B) The speed of the reaction is slowed by the ingestion of more starches.
(C) The amount of substrate is limited.
(D) All enzymes will aid in the catalysis of starches.
(E) Bacteria is not present at the site.

79. Which of the following is an autotroph?

(A) *E. coli* bacteria
(B) Portuguese man-of-war jellyfish
(C) Portobello mushroom
(D) Asparagus fern
(E) A human fetus

80. A _____ is a distinct group of individuals that are able to mate and produce viable offspring.

(A) class
(B) community
(C) phylum
(D) family
(E) species

81. Which phylum of vertebrates contains animals with mammary glands?

(A) Chondrichthyes
(B) Amphibia
(C) Reptilia
(D) Mammalia
(E) Aves

Questions 82–86 consist of five lettered terms followed by a list of numbered phrases. For each numbered phrase, select the one term that is most closely related to it. Each term may be used once, more than once, or not at all.

(A) Natality
(B) J-curve
(C) Population rate of growth
(D) Mortality
(E) S-curve

82. Exponential population growth curve

83. Birthrate minus death rate

84. Death rate within a population

85. Birthrate within a population

86. Logistic population growth

87. Which of the following kingdoms contains photosynthetic organisms?

 (A) Fungi
 (B) Protista
 (C) Animalia
 (D) Monera
 (E) Mycota

88. Which of the following elements is generally NOT found in organic tissue?

 (A) Oxygen
 (B) Hydrogen
 (C) Nitrogen
 (D) Argon
 (E) Sulfur

Questions 89–93 consist of five lettered terms followed by a list of numbered phrases. For each numbered phrase, select the one term that is most closely related to it. Each term may be used once, more than once, or not at all.

 (A) Cuticle
 (B) Nonvascular plants
 (C) Angiosperms
 (D) Gymnosperms
 (E) Lateral buds

89. Bryophytes

90. Flowering plants

91. Waxy coating of leaves that maintains moisture balance

92. Produce seeds without flowers

93. Conifers and cycads

94. Organisms store energy within

 (A) carbon
 (B) chemical bonds
 (C) hydrogen
 (D) water
 (E) phosphorous

95. The diploid generation in plants is known as the

    (A) prothallus
    (B) sporophyte
    (C) gametophyte
    (D) adult
    (E) spore

96. There are three different genes that control skin color in humans. Each gene has a dominant and a recessive allele, so the possible alleles are A, a, B, b, C, c. The more dominant alleles inherited by offspring, the darker the skin color. Skin color in humans is an example of

    (A) a polygenic trait
    (B) an autosome
    (C) a sex-linked trait
    (D) a monohybrid cross
    (E) a dihybrid cross

97. The fossil of a fish is found in a limestone bed. The imprint of a skeleton is easily discernable, including several vertebrae. The fish most likely belonged to the class

    (A) Cephalochordates
    (B) Porifera
    (C) Osteichthyes
    (D) Cnidaria
    (E) Platyhelminthes

98. Cells of eukaryotes have all of the following EXCEPT

    (A) membrane-bound organelles
    (B) DNA organized into chromosomes
    (C) a nucleus
    (D) DNA floating free in the cytoplasm
    (E) ribosomes

99. The process whereby molecules and ions flow through a cell membrane from an area of higher concentration to an area of lower concentration without an input of energy is known as

(A) diffusion
(B) active transport
(C) endocytosis
(D) exocytosis
(E) phagocytosis

100. A species' ecological niche is defined as including

(A) only the physical features of its habitat
(B) all the biotic and abiotic factors that will support its life and reproduction
(C) the biotic features of its habitat
(D) only abiotic factors such as weather, temperature, etc.
(E) only its place in the food chain

101. A DNA strand in a double helix has a base sequence of ATACGT. The base sequence of its DNA complement is

(A) ACGUAU
(B) ATACGT
(C) TATGCA
(D) TGCATA
(E) UAUGCA

102. According to the Hardy-Weinberg Law, evolution can occur due to changes in allele frequencies. All of the following can contribute to changes in allele frequencies EXCEPT

(A) mutations
(B) immigration
(C) emigration
(D) natural selection
(E) sexual recombination

103. There are various types of plant stems that have different functions. All of the following are types of stem EXCEPT

   (A) tendrils
   (B) nodes
   (C) tubers
   (D) rhizomes
   (E) corms

Questions 104–106 refer to the diagram of the organelle below.

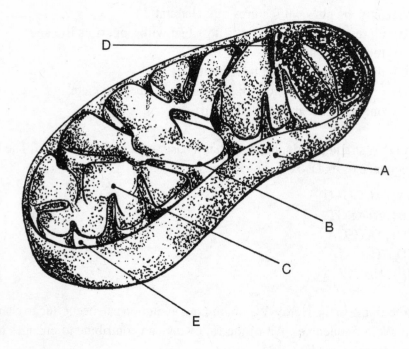

104. The Krebs Cycle (citric acid cycle) occurs here.

105. The majority of this organelle's energy harvest occurs here.

106. Anaerobic metabolism of glucose occurs here.

107. Protists are divided into two major subgroups by their

    (A) methods of locomotion
    (B) methods of reproduction
    (C) habitats
    (D) chromosome numbers
    (E) methods of nutrition

108. A gas that causes asphyxiation by binding to hemoglobin, thus preventing oxygen from doing so, is

    (A) carbon dioxide
    (B) carbon monoxide
    (C) nitrous oxide
    (D) sulfer dioxide
    (E) water vapor

109. Regarding the taxonomic classification of man,

    (A) its phylum is Mammalia
    (B) its family is Hominidae
    (C) its genus name is *sapiens*
    (D) its kingdom is Chordata
    (E) its order is Vertebrata

110. The diversification of mammals that followed the extinction of dinosaurs is an example of

    (A) allopatric speciation
    (B) sympatric speciation
    (C) disruptive selection
    (D) adaptive radiation
    (E) natural selection

Questions 111–114 refer to stages of meiosis I. Choose the capital letter on the illustration that most directly applies to the word seen in the question.

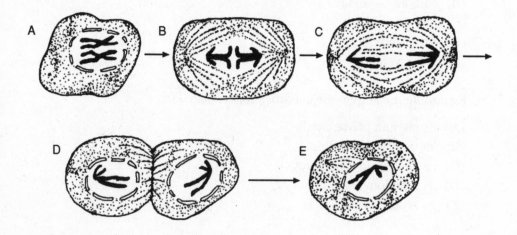

111. anaphase

112. metaphase

113. prophase

114. telophase

115. Hydrolysis of lipid molecules yields:

    (A) amino acids and water
    (B) amino acids and glucose
    (C) fatty acids and glycerol
    (D) glucose and glycerol
    (E) glycerol and water

PRACTICE TEST 1

Answer Key

1.	(B)	39.	(A)	77.	(B)
2.	(C)	40.	(B)	78.	(A)
3.	(C)	41.	(C)	79.	(D)
4.	(E)	42.	(E)	80.	(E)
5.	(A)	43.	(C)	81.	(D)
6.	(E)	44.	(E)	82.	(B)
7.	(B)	45.	(E)	83.	(C)
8.	(A)	46.	(A)	84.	(D)
9.	(B)	47.	(B)	85.	(A)
10.	(D)	48.	(A)	86.	(E)
11.	(E)	49.	(D)	87.	(B)
12.	(D)	50.	(D)	88.	(D)
13.	(A)	51.	(A)	89.	(B)
14.	(C)	52.	(C)	90.	(C)
15.	(A)	53.	(C)	91.	(A)
16.	(A)	54.	(A)	92.	(D)
17.	(A)	55.	(B)	93.	(D)
18.	(A)	56.	(E)	94.	(B)
19.	(D)	57.	(B)	95.	(B)
20.	(E)	58.	(A)	96.	(A)
21.	(E)	59.	(D)	97.	(C)
22.	(A)	60.	(C)	98.	(D)
23.	(B)	61.	(D)	99.	(A)
24.	(A)	62.	(C)	100.	(B)
25.	(A)	63.	(D)	101.	(C)
26.	(D)	64.	(B)	102.	(E)
27.	(E)	65.	(A)	103.	(B)
28.	(A)	66.	(B)	104.	(C)
29.	(B)	67.	(B)	105.	(B)
30.	(C)	68.	(B)	106.	(D)
31.	(A)	69.	(C)	107.	(E)
32.	(D)	70.	(E)	108.	(B)
33.	(A)	71.	(A)	109.	(B)
34.	(B)	72.	(B)	110.	(D)
35.	(C)	73.	(D)	111.	(C)
36.	(A)	74.	(E)	112.	(B)
37.	(D)	75.	(A)	113.	(A)
38.	(E)	76.	(C)	114.	(D)
				115.	(C)

PRACTICE TEST 1

Detailed Explanations of Answers

1. **(B)** Inorganic cofactors are substances that promote enzyme catalysis. These molecules may bind to the active site or to the substrate itself. The most common inorganic cofactors are metallic elements such as iron, copper, and zinc. Coenzymes are organic in nature, thus they are organic cofactors. Hormones are also organic and are not cofactors. Similarly, inorganic cofactors are minerals, but not all minerals are inorganic cofactors.

2. **(C)** The filament, anther, pollen grains, and tube nuclei are all parts of the male reproductive system in plants. The male structure is the stamen, consisting of the anther atop a long, hollow filament. The anther has four lobes and contains the cells that become pollen. The tube nuclei develop from pollen grains. The stigma is only found on female plants.

3. **(C)** Blue-green bacteria are prokaryotic organisms of the Kingdom Monera. Prokaryotes do not have a nucleus or any other membrane-bound organelles. The cells of mold, seaweed, ferns, and hydra all have membrane-bound organelles and are therefore classified as eukaryotes.

4. **(E)** Scientists conclude that in the evolution of life (including plants) the first cells were prokaryotic, and that eukaryotic cells developed as cells with varying functions that were incorporated into more complex cells (the Endosymbiont Theory). In addition the early Earth atmosphere was void of oxygen, so early cells were anaerobic; aerobic cells evolved later. Cyanobacteria are autotrophic (photosynthetic) prokaryotes that are considered ancestors of multicellular plants. Both answers (A) and (B) incorrectly place eukaryotes before prokaryotes. Answer (C) incorrectly places aerobic cells before anaerobic cells. Answer (D) incorrectly places anaerobic cells between aerobic cells and plants.

5. **(A)** Types of nitrogen-fixing bacteria live in symbiosis in the nodules of the roots of legumes (beans, peas, clover, etc.) supplying the roots with a direct source of ammonia—nitrogen in a form usable by plants. This is a unique niche (changing nitrogen from an unusable to a usable form) filled by the symbiotic relationship of legumes with nitrogen-fixing bacteria.

6. **(E)**   A carbohydrate molecule contains only carbon, hydrogen, and oxygen in the ratio CH_2O. Nitrogen, N, and sulfur, S, are elements that are not found in carbohydrates. The amine, NH_2, and phosphate, PO_4, groups are found in other organic molecules, but not in carbohydrates.

7. **(B)**   The hydrologic cycle intersects with nearly all biogeochemical cycles. The cycle includes the evaporation from bodies of water and plant leaves; water vapor is then redistributed over land via clouds that release water as precipitation. Water flows back into waterways that eventually flow into the ocean. Water vapor in the air is a greenhouse gas that traps heat in the lower atmosphere and resists rapid cooling.

8. **(A)**   Glucose is the only monosaccharide listed. Cellulose and starch are both polysaccharides. Table sugar is a disaccharide of glucose and fructose (the most common monosaccharides). Amylase is not a sugar at all. It is an enzyme important to the digestion of starches.

9. **(B)**   Innate behaviors are the actions in animals we call instincts. Innate behaviors are highly stereotyped; they are similar actions performed by many individuals.

10. **(D)**   A reflex is an automatic movement of a body part in response to an environmental signal.

11. **(E)**   A fixed action pattern (FAP) is the most complex of stereotyped behaviors. It is a pre-programmed and complex response to a particular environmental signal known as a *releaser* or a *sign stimulus*. FAPs include courtship behaviors, circadian rhythms, and feeding of young. Organisms automatically perform FAPs without any prior experience (they are not learned).

12. **(D)**   Ecology is literally "the study of" (*-ology*) a "place to live" (*eco*). It involves the study of all the living (including interaction with other organisms) and nonliving factors that contribute to an organism's life within its living space.

13. **(A)**   An enzyme is a special protein that acts as a catalyst for organic reactions. A catalyst is a substance that changes the speed of a reaction without being affected itself. The enzyme will speed up the initial reaction rate until all the enzymes are in use, and then the reaction rate will level off as all the limited amount of enzyme becomes engaged but substrate is still present. Enzymes will not decrease the reaction rate or stop the reaction completely. Furthermore, the reaction will not stop occurring when all the enzyme is

engaged, but the reaction rate will level off. The rate cannot continue to rise after all the enzyme is engaged.

14. **(C)**   The spinal cord is not directly involved with immunity. The organs of the immune system in humans and other higher invertebrates include all of the other organs listed in the question—the lymph nodes, spleen, thymus, and tonsils.

15. **(A)**   The reaction shown is a common combination reaction in metabolism that stores energy within the chemical bonds of triglyceride (fat) molecules. Therefore, the reaction is endothermic (energy storing), not exothermic (energy releasing). This reaction is a common part of animal energy storage. Water is produced by this reaction.

16. **(A)**   The term *prokaryote* is derived from "pro" (meaning "before") and "karyo" (meaning "a nucleus"); thus the literal definition is "before nucleus." Prokaryotic cells have no nucleus or any other membrane-bound organelles. The DNA in prokaryotic cells floats freely within the cytoplasm. A eukaryotic cell, by contrast, contains a nucleus and other membrane-bound organelles. *Endocytic* refers to the ability of a cell to take in substances from its environment, not to the existence or lack of membrane-bound organelles.

17. **(A)**   Fungi are not photosynthetic—they absorb nutrients from their surroundings. Therefore, fungi cells do not have chloroplasts. The cells of fungi are eukaryotic and, therefore, do have a nucleus, mitochondria, plasma membranes, and ribosomes.

18. **(A)**   Grana are embedded within the stroma (or body) of the chloroplasts in plant cells. Pigments, including chlorophyll and carotene (when present), are found within the grana. Cristae are found in mitochondria, not in chloroplasts.

19. **(D)**   Nearly all life forms derive their energy from the Sun whether directly (through photosynthesis) or indirectly (by consuming other organisms).

20. **(E)**   The process of photosynthesis is the crucial reaction that converts the light energy of the Sun into chemical energy that is usable by living things. While catabolism, metabolism, oxidative phosphorylation, and protein synthesis are necessary processes for most organisms, only photosynthetic organisms convert energy directly from light.

21. **(E)** The sum total of genetic information of a species is its genome. Genes are portions of chromosomes that determine the inheritance of an organism's characteristics. Polyploidy refers to the existence of a multiple of the normal number of chromosomes in a cell.

22. **(A)** Free bases line up along the DNA template and are bonded together forming RNA during transcription, not translation. Translation begins as a ribosome attaches to the mRNA strand at a particular codon known as the start codon. This codon is only recognized by a particular initiator tRNA. The ribosome continues to link tRNA whose anticodons complement the next codon on the mRNA string. A third type of RNA is utilized at this point, ribosomal RNA or rRNA. Ribosomal RNA exists in concert with enzymes as a ribosome. In order for the tRNA and mRNA to link up, enzymes connected to rRNA at the ribosome must be utilized. Ribosomal enzymes also are responsible for linking the sequential amino acids into a protein chain. As the protein chain is forming, the ribosome moves along the sequence adding the amino acids that are designated by the codons on the mRNA. At the end of the translation process, a terminating codon stops the synthesis process and the protein is released. Thus, answers B–E represent steps in the translation process.

23. **(B)** Transduction refers to the changing of a bacterium's genetic makeup by the transfer of a portion of genetic material from one bacterial cell to another. This is accomplished via a bacteriophage (a virus that targets bacteria). Transformation refers to the absorption and incorporation of pieces of DNA from a bacterium's environment (usually from dead bacterial cells).

24. **(A)** Chromosomes that are paired with others of similar size and shape within the nucleus are known as homologous pairs or homologs. Each set of homologous chromosomes has a similar genetic constitution, but the genes are not necessarily the same.

25. **(A)** Restriction enzymes cut out sections of DNA molecules by cleaving the sugar-phosphate backbone. In most cases, a particular enzyme cuts the strands of the double helix DNA within a length of a few bases.

26. **(D)** Telophase follows anaphase. As the chromatids arrive at opposite poles of the cell, nuclear membranes form around the new groups of single-stranded chromosomes. Chromosomes disperse through the new nucleoplasm and are no longer visible as chromosomes. Spindle fibers also disappear in telophase.

27. **(E)** Telophase ends with the splitting of the cytoplasm into two distinct cells, a process known as cytokinesis.

28. **(A)** During prophase, the first stage of mitosis, the chromatin condenses into chromosomes; the centrioles move to opposite ends of the cell, and spindle fibers begin to extend from the centromeres of each chromosome toward the center of the cell. In the second part of prophase the nuclear membrane dissolves and the spindle fibers attach to the centromeres at the kinetochore.

29. **(B)** During metaphase the spindle fibers pull the chromosomes into alignment along the equatorial plane of the cell, ensuring that one copy of each chromosome is distributed to each daughter cell.

30. **(C)** Anaphase is characterized by the separation of the paired chromosomes at the kinetochore. Each chromosome travels along the spindle fibers to opposite ends of the cell.

31. **(A)** Mitosis produces two exact copies of the original cell as daughter cells. Meiosis, on the other hand, is the process of producing four daughter cells that have single chromosomes (haploid). The parent cell is diploid, that is, it has sets of chromosomes. Meiosis also results in the production of four new cells, rather than two (as in mitosis), and each cell has half the chromosomes of the parent. Meiosis occurs in reproductive organs. Meiosis, like mitosis, begins with chromosome duplication. In meiosis, two distinct nuclear divisions occur: the first is known as reduction (or meiosis 1) and the second, division (or meiosis 2). Reduction affects the diploid level reducing it from 2n to n (i.e., diploid to haploid). Division then distributes the remaining chromosomes in a mitosis-like process among four daughter cells.

32. **(D)** Oxygen is carried by hemoglobin molecules (that contain iron) in red blood cells. Lymphocytes, erythrocytes, eosinophils, and neutrophils are all non-oxygen-carrying types of blood cells.

33. **(A)** Osmosis is a type of diffusion (passive transport) that occurs only with water molecules. The water on the side of the membrane with the highest water pressure will cross through the membrane until the concentration is equalized on both sides. Facilitated diffusion is the movement of substances across the cell membrane with the help of specialized proteins. Active transport is the movement of substances across membranes with the help of added energy. Exocytosis is the process where large molecules

are engulfed in a pocket of cell membrane and are exported from the cell. Endocytosis also involves large molecules that are engulfed in a pocket of membrane, but in this case the molecules are imported into the cell.

34. **(B)** The epiglottis is the flap of tissue that covers the glottis, preventing food particles from entering the bronchial tubes. Before reaching the glottis (the opening that allows gases to pass into the bronchi), air passes through the pharynx and into the trachea. The larynx is the upper portion of the trachea; the glottis is the lower portion of the trachea.

35. **(C)** The placenta is the connection between the mother and embryo; it is the site of transfer for nutrients, water, and wastes between them. The yolk sac stores nutrients for the embryo and the yolk sac membrane encloses the yolk sac. The uterus is the organ in the female body that houses the developing embryo. A portion of the uterine lining becomes a part of the placenta, but the entire uterus is not involved. The allantois, contained by the allantoic membrane, develops into part of the umbilical cord. The gastrula is the embryo at a stage of cell division that forms into a double-layered tube.

36. **(A)** Digestive enzymes are released by the pancreas and gall bladder into the small intestine. Digested food then passes from the small intestine to the large intestine. Food travels through the esophagus on its way to the stomach, where food is digested through mechanical means as well as chemical means, but not with enzymes released by the pancreas and gall bladder. The stomach produces its own enzymes, mucus, and gastric juices. The rectum connects the large intestine with the anus where waste products are transported out of the body.

37. **(D)** The heart is made of cardiac muscle. The electrical properties of cardiac muscle tissue cause the beating of the heart muscle which results in the pumping of blood through the body. When a heart stops beating, stimulating the cardiac muscle with electric shock can sometimes restart it. Smooth muscle tissue is not found in the heart.

38. **(E)** All the DNA of every cell of every organism is copied repeatedly to form new cells for growth, repair, and reproduction. A mutation can be an error that occurs randomly during one of the many copying sequences that occur within the cell. Mutations can also be the result of damage to DNA through environmental agents such as sunlight, cigarette smoke, chemical exposure, or x-rays. Crossing over is a process that may occur during

meiosis, resulting in exchange of corresponding portions between homologous chromosomes. Crossing over is not a mutation.

39. **(A)** Primary oocytes are formed in the ovaries of females before birth, usually in great number.

40. **(B)** The egg cell is the female gamete (haploid) ready for fertilization. Primary oocytes go through meiosis I to become a secondary oocyte and a polar body. The polar body goes through meiosis II forming two more polar bodies, and the secondary oocyte then goes through meiosis II to form one egg cell and one polar body.

41. **(C)** Primary spermatocytes undergo meiosis I to form secondary spermatocytes with a single chromosome set (haploid). These secondary spermatocytes then go through meiosis II, forming spermatid, which are haploid. Spermatids develop into male gametes (sperm cells).

42. **(E)** Three polar bodies are formed when primary oocytes undergo meiosis I and II. The fourth cell that is formed is an egg cell. Polar bodies are not fertile.

43. **(C)** See answer to question #41.

44. **(E)** In addition to being the primary structure and support for the human frame, the 206 bones of the skeleton protect the soft internal organs of the human body, produce red blood cells from its marrow, and allow for movement by providing a base for muscles and ligaments. While the bones do store calcium and phosphates, they do not produce them.

45. **(E)** The illustration is called a Punnett square and is used to illustrate genetic crosses. The genotype is the representation of the alleles present in a particular organism (for example, Tt, where the uppercase T stands for the dominant allele for "tall" and the lowercase t stands for the recessive allele for short), while the phenotype is the description of the physical attribute of a particular genotype (for example, the phenotype for Tt would be "tall" since tall is dominant). While Mendel was instrumental in the study of genetics, Reginald Punnett, not Mendel, developed the Punnett square; thus, it is not called a "Mendelian Diagram." *Phenogram* is not a term that is used in genetics.

46. **(A)** Since the genotypes of the parents are always listed on the side and top of a Punnett square, the genotypes of the parents are TT and tt.

47. **(B)** Every gamete will have either the T allele or the t allele since those are the only possibilities of gametes from the parents.

48. **(A)** All offspring will have a genotype Tt; all have a dominant gene for tallness (T), so all will have a phenotype of tall; thus, the ratio is 4 tall: 0 short.

49. **(D)** The Punnett square for a cross of two of the offspring would be:

	T	t
T	TT	Tt
t	Tt	tt

The genotype ratio will be 1 TT : 2 Tt : 1 tt. Since T (tall) is dominant to t (short), only one out of four will be short. The phenotypic ratio will be 3 tall : 1 short.

50. **(D)** There is no epistematic region in plant roots. Roots have four major structural regions that run vertically from bottom to top. The root cap is composed of dead, thick-walled cells and covers the tip of the root, protecting it as the root pushes through soil. The meristematic region is just above the root cap. It consists of undifferentiated cells that carry on mitosis, producing the cells that grow to form the elongation region. In the elongation region, cells differentiate, large vacuoles are formed, and cells grow. As the cells differentiate into various root tissues, they become part of the maturation region.

51. **(A)** The hydrogen atoms in water molecules have a partial positive charge and the oxygen atoms a partial negative charge, causing polarity. This polarity allows the oxygen of one water molecule to attract the hydrogen of another. The partial charges attract other opposite partial charges of other water molecules, allowing for weak (hydrogen) bonds between the molecules. "Inert" means non-reactive; it does not explain the attraction between H and O. There are no ionic bonds within water molecules, only covalent bonds. Crystal structure forms in ice because of the attraction of hydrogen bonds; the crystal structure does not cause the attraction. Brownian motion is the random movement of atoms or particles caused by collisions between them; it does not explain attraction between atoms or molecules.

52. **(C)** The first law of Mendelian genetics is the Law of Segregation, stating that traits are expressed from a pair of genes in the individual, one of which came from each parent. The alleles are randomly separated as gametes are formed and are brought together in varying combinations through fertilization.

   Not all genetic crosses involve one allele that is dominant over the other and not all crosses involve multiple alleles. Also, mutations may occur during the DNA replication process of gamete formation; however, gamete formation itself does not cause mutation. Offspring do not always have the same genotype for a trait as a parent; rather alleles are independently sorted to give the offspring a combination of alleles with one from each parent.

53. **(C)** Animal species are capable of sexual reproduction, though some, such as the hydra and other invertebrates, reproduce asexually. All animal cells are eukaryotic (have nuclei and membrane-bound organelles). Only plants and blue-green bacteria have cell walls and/or plastids. Animals do develop from embryos and are heterotrophic, which means that they gain their energy directly or indirectly by ingesting autotrophs.

54. **(A)** Competition occurs when niches overlap between two species in the same community. The term "competition" does not indicate movement in or out of an area. Emigration (one-way movement out of the original range), immigration (one-way movement into a new range), and migration (temporary movement out of a range and then back to it) are all forms of dispersion (movement of species).

55. **(B)** The similarity of body shapes between sharks and dolphins most likely results from the convergent evolution of traits that are favorable to survival in the ocean. Sharks (cartilaginous fish) and dolphins (mammals) do not share a common ancestry, nor did one evolve into the other. Sharks and dolphins have different niches in the ocean community, so they are not in competition. Allopatric speciation occurs between geographically isolated populations of the same species that eventually develop into separate species.

56. **(E)** The center of the root contains the vascular cylinder (e), including xylem and phloem tissue.

57. **(B)** The outermost layer of primary tissue is the epidermis (b). The epidermis is one cell layer thick. It protects the internal root tissue and absorbs nutrients and water.

58. **(A)** In the maturation region the epidermis produces root hairs (a), extensions of the cells that reach between soil particles and retrieve water and minerals.

59. **(D)** Inside the cortex is a ring of endodermis (d), a single layer of cells that are tightly connected so no substances can pass between cells. This feature allows the endodermis to act as a filter; all substances entering the vascular tissues from the root must pass through these cells.

60. **(C)** The ring inside the epidermis is known as the cortex (c), made of large parenchyma cells. Parenchyma cells are thin-walled cells loosely packed to allow for flow of gases and mineral uptake.

61. **(D)** The term *Homo sapiens* is an example of binomial nomenclature, the use of the genus and species names together.

62. **(C)** According to the Hardy-Weinberg Law, $p + q = 1$, where p and q represent the frequencies of two alleles. Also, $p^2 + 2pq + q^2 = 1$, where the frequency of homozygous dominant genotypes is represented by p^2, the homozygous recessive by q^2, and the heterozygous genotype by 2pq. So, in the case of tongue-rolling, if the frequency of non-tongue-rollers is 36% or 0.36, then the frequency of the recessive allele (q) is 0.6 (the square root of 0.36). Therefore, the frequency of the dominant allele (p) is 0.4 (since $p + q = 1$). The heterozygous genotype is represented by 2pq, which equals 2(0.6)(0.4) or 0.48.

63. **(D)** Five petals indicates that the plant is a dicot (a plant with two cotyledons in each seed). Dicots also have taproot systems and leaves with networked or branching veins. The vascular bundles of a dicot will be arranged in rings. A monocot would have random arrangement of its vascular bundles and leaves with parallel veins. There was no indication from the given information whether the plant was male or female.

64. **(B)** The order of classification from least specific to most specific is kingdom, phylum, class, order, family, genus, and species; so of those listed, genus is the most specific, with members most closely related to each other.

65. **(A)** Oparin's hypothesis included the idea that most water on Earth was in the form of water vapor and steam, not ice. Answer choices (B)–(E) are all consistent with Oparin's hypothesis, which proposed that the Earth was approximately 4.6 billion years old and had a reducing atmosphere with very little oxygen present. There was an abundance of ammonia, hydrogen, methane, water vapor, and steam (H_2O). There was a great deal of

heat energy available as the Earth was cooling. Recurring violent lightning storms also provided energy. The cooling of the Earth also caused much of the water vapor surrounding the Earth to condense, forming hot seas. The concentration of organic molecules became very high; they began forming into larger, charged, complex molecules called coacervates.

66. **(B)** Though the presence of altruistic traits may actually decrease the individual's fitness to survive, it increases the survival rate of the population. Altruistic traits are preserved through kin selection. Kin selection accounts for the expression of altruistic traits toward close relatives, thus increasing the probability for those relatives to survive and to pass on their genetic traits.

67. **(B)** Natural selection occurs when the surviving species is the one that is most adapted to the environment (in this case, resistant to penicillin). Clearly, only the surviving competitors reproduce. Therefore, traits that provide the competitive edge will be represented most often in succeeding generations. The survival of resistant bacteria does not represent genetic drift. Genetic drift is random fluctuation in allele frequency, including loss of alleles. It is most pronounced in small populations and therefore in populations that have become separated from a main population. In this example, mutation does not contribute to the survival of resistant bacteria; the genetic material has not changed. There is no genetic equilibrium, since a certain trait (resistance) is favored by the environment. Allopatric speciation depends on the physical isolation of two portions of a population, and produces entirely new species over time.

68. **(B)** The driving mechanism of evolution is natural selection. The Oparin hypothesis involved only the earliest forms of organic life and does not involve speciation. Allopatric speciation and genetic drift both depend upon natural selection. Hardy-Weinberg Equilibrium explains the preservation of genes within a population in spite of the occurrence of natural selection.

69. **(C)** Punctuated equilibrium is a model that proposes that species undergo a long period of equilibrium, then the equilibrium is upset by environmental forces, causing a short period of quick mutation and change, resulting in the relatively rapid evolution of new species.

70. **(E)** Genetic drift occurs as a gene pool experiences a change in frequency of particular genes due to chance fluctuations. Over time the genetic pool within this finite population changes.

71. **(A)** When populations of an organism in a given area grow and move into new geographic areas, some discover new niches and advantageous conditions. Traits possessed by this traveling population will grow more common over several generations through the process of natural selection. Over time the species will specially adapt to live more effectively in the new environment. Through this process known as adaptive radiation, a single species can develop into several diverse species over time.

72. **(B)** A geographic separation within a population may develop members with a genetic difference that prevents successful reproduction with the original species. The genetically different members reproduce with each other, producing a population that is separate from the original species. This process is called sympatric speciation.

73. **(D)** Kin selection results in the preservation of traits that are advantageous to a population because organisms are more likely to exhibit altruistic traits toward close relatives. The altruistic traits contribute to the survival of close relatives. Close relatives will share more genetic information, so the traits of the relatives are more likely to be preserved, since they are the survivors. The preserved traits include the altruistic traits and therefore are likely to be expressed in succeeding generations.

74. **(E)** Dispersal of species to an island is dependent on geographic as well as historical factors. Distance from other landmasses, prevailing winds, and ocean currents are also geographic factors that will affect species introduction. Historical factors such as climate shifts (for instance, the shift to an ice age), drought, volcanic action, plate shifting, and human intervention will also affect what species are able to travel to a given island. Fossil preservation or lack thereof would not affect any current conditions of biogeography.

75. **(A)** Parasitism is symbiosis in which one organism benefits, but the other is harmed. Mutualism is symbiosis that benefits both organisms. Amensalism is symbiosis where one organism is neither helped nor harmed but the growth of the other is inhibited. Predation is not symbiosis; rather it is the killing of an organism by another. Habituation is also not symbiosis; it is a behavioral response in which there is less and less response by an individual to a stimulus over time.

76. **(C)** Animal respiration releases carbon dioxide back into the atmosphere in large quantities; it does not take in carbon dioxide for use. Most of the

carbon within organisms is derived from the production of carbohydrates in plants through photosynthesis. Carbon dioxide is dissolved directly into the oceans, where it is combined with calcium to form calcium carbonate—used by mollusks to form their shells. Detritus feeders include worms, mites, insects, and crustaceans that feed on dead organic matter, returning the carbon to the cycle through chemical breakdown and respiration. Organic matter that is left to decay may under conditions of heat and pressure be transformed into coal, oil, or natural gas—the fossil fuels. When fossil fuels are burned for energy, the combustion process releases carbon dioxide back into the atmosphere, where it is available to plants for photosynthesis.

77. **(B)** Large molecules (such as proteins) are not able to pass through the cell membrane, but are instead engulfed by the cell membrane. Endocytosis is the process whereby large molecules (i.e., some sugars or proteins) are taken up by a sack of membrane and delivered to the interior of the cell where they can be used. This process, for instance, is used by white blood cells to engulf bacteria. Exocytosis uses the same processes but exports substances to the exterior of the cell. Diffusion, active transport, and facilitated diffusion all refer to processes of passing substances through the cell membrane by various means.

78. **(A)** It is the shape of the active site of the enzyme that allows the enzyme to bind to the substrate to form product(s). The active site of the amylase molecule matches the shape of starch molecules. The speed of the reaction and the amount of substrate do not enhance amylase's enzymatic functions. Specific enzymes are needed for specific reactions.

79. **(D)** Plants (including asparagus ferns) produce their own food through photosynthesis and are known as autotrophs. Fungi, jellyfish, human fetuses, and bacteria do not produce their own food.

80. **(E)** By definition, a species is a distinct group of individuals that are able to mate and produce viable offspring. Class, phylum, and family are taxonomic groups. A community includes all the species that interact in a particular area.

81. **(D)** Only mammals (mammalia) have mammary glands.

82. **(B)** Exponential population growth is represented by the J-curve. The rate of growth accelerates over time since there are no limiters of growth.

83. **(C)** The rate of increase within a population is represented by the birth rate minus the death rate.

84. **(D)** Mortality is the death rate within a population.

85. **(A)** Natality is the birth rate within a population.

86. **(E)** Logistic population growth is represented by the S-curve. It occurs in populations that encounter limiting factors where acceleration occurs to a point then slows down.

87. **(B)** Photosynthetic organisms are found in the Kingdom Plantae and the Kingdom Protista, and are not found in the Kingdom Fungi, Kingdom Animalia, or Kingdom Monera. *Mycota* is another term for Fungi.

88. **(D)** Argon is a noble gas; it does not occur naturally in organic tissue. All organic molecules contain carbon, but also commonly contain oxygen, hydrogen, nitrogen, and sulfur (as well as phosphorous, not listed in answers).

89. **(B)** Nonvascular plants are known as bryophytes (e.g., mosses). They lack tissue that will conduct water or food.

90. **(C)** Angiosperms are those plants that produce flowers as reproductive organs.

91. **(A)** The waxy coating on leaves that keeps moisture balanced is the cuticle.

92. **(D)** Gymnosperms produce seeds without flowers.

93. **(D)** Gymnosperms produce seeds without flowers, and include conifers (cone-bearers) and cycads.

94. **(B)** Chemical bonds are where energy is stored within cells.

95. **(B)** The diploid (2n) generation in plants is known as the sporophyte. The prothallus is the haploid (n) structure that develops into the mature gametophyte in ferns. The gametophyte is always haploid as well. What we consider the adult generation in a plant's life cycle may be haploid (e.g., mosses) or diploid (e.g., ferns). Spores are male haploid gametes.

96. **(A)** When more than one gene controls a particular genetic trait, that trait is called polygenic. An autosome is any chromosome that does not determine the sex of an individual. A sex-linked trait is one whose genes are

found on one of the sex chromosomes. A monohybrid cross is a genetic cross where only one trait is considered. A dihybrid cross considers two traits.

97. **(C)** Since the skeleton of the fish was easily identifiable, the fish must belong to the class of bony fish—Osteichthyes. Cephalochordates have a notochord, but no vertebrae. Porifera is the phylum including sponges; Cnidaria is the phylum including jellyfish and hydra; Platyhelminthes is the phylum containing flat worms.

98. **(D)** The DNA of eukaryotes is organized into chromosomes within the nucleus.

99. **(A)** Diffusion is the process whereby molecules and ions flow through the cell membrane from an area of higher concentration to an area of lower concentration. Where the substance exists in higher concentration, collisions occur that tend to propel them away toward lower concentrations. Active transport requires added energy to move substances across a membrane. In endocytosis and exocytosis the cell membrane surrounds the substance and moves it either into (*endo*) or out of (*exo*) a cell. Phagocytosis occurs when a portion of a cell's membrane engulfs and destroys a foreign body.

100. **(B)** A species' habitat must include all the factors that will support its life and reproduction including biotic factors (i.e., living—food source, predators, place in the food chain, etc.) and abiotic (i.e., nonliving—weather, temperature, soil features, etc.).

101. **(C)** Two base-pairing rules must be memorized for DNA strands: A–T and C–G. Thus, the given DNA strand of six bases dictates only one possible complement.

102. **(E)** Hardy-Weinberg Equilibrium only occurs when mutation, immigration, emigration, and natural selection are not occurring in a population. When mutation, immigration, emigration, and natural selection are occurring, evolution is possible as allele possibilities are changed. Sexual recombination is a factor that will reinforce Hardy-Weinberg Equilibrium.

103. **(B)** There are many types of stems that have specialized functions. The functions of stems generally may include transport of water and food between root and leaves, leaf support, and food storage. Tendrils are modified stems that assist climbing plants such as the grapevine. Underground

stems include tubers, rhizomes, and corms. Tubers, found in the potato, function to store starch. Rhizomes, in ferns, function in vegetative propagation. Corms are found in gladiolus and are actually fleshy leaves that store food. Nodes are not stems, but rather are the site on the stem at which the leaves attach. Internodes are thus the region between nodes.

104. **(C)** This organelle is the mitochondrion, the site of cellular respiration. Pyruvic acid enters the Krebs cycle (citric acid cycle) here, where energy released by oxidation reactions performed on pyruvic acid is stored in high-energy phosphate bonds of ATP. ATP, adenosine triphosphate, is the molecule used by all cells to store energy.

105. **(B)** The inner surface of a mitochondrion's inner membrane has bound to it molecules that take part in oxidative phosphorylation. These molecules are alternatively reduced and oxidized, which releases energy that can be used to synthesize ATP.

106. **(D)** The anaerobic metabolism of glucose is known as glycolysis. Unlike the aerobic phase of metabolism in the mitochondrion, the team of enzymes running the glycolytic pathway are found in the cytoplasm near the mitochondrion.

107. **(E)** Members of the kingdom Protista may be autotrophic, heterotrophic, or a combination of both depending on the presence or absence of chloroplasts. This criterion can be applied to the three types of protists: algae, slime molds, and protozoa.

108. **(B)** Carbon monoxide is a colorless, odorless gas that can bind to hemoglobin without the subject's awareness. As hemoglobin in red blood cells transports oxygen, oxygen's unavailability due to CO binding causes internal suffocation, or asphyxiation.

109. **(B)** Taxonomy refers to the scientific classification of all living things into a systematic scheme. It is based on evolutionary relationships. The largest category of classification is the kingdom, of which there are five: Monera (bacteria and blue-green algae), Protista, Fungi, Plantae, and Animalia. Within a kingdom, there are phyla (or divisions as in Fungi and Plantae). Phyla are divided into classes, classes into orders, and orders into families. Families are divided into genera. Each genus is then divided into species. When we refer to a specific organism, we usually give its binomial name, the genus and species. Hence, man is called *Homo sapiens*. Note that

by convention, the genus name is capitalized and the specific epithet is not; the binomial name is underlined or italicized.

The specific epithet NEVER stands alone. Thus, man's species name is *Homo sapiens,* not *sapiens*. The specific name could be given to organisms of other genera. For example, *multiflora* is a specific name. When used alone, you could be referring to *Rosa* or *Begonia*, clearly two different species. By saying *Rosa multiflora*, you are referring to one and only one species.

Man's complete taxonomic classification is as follows:

kingdom	—	Animalia
phylum	—	Chordata
class	—	Mammalia
order	—	Primates
family	—	Hominidae
genus	—	*Homo*
species	—	*Homo sapiens*

There are also sub-groups. For instance, man belongs in the subphylum Vertebrata. For other species, there may be subclasses, suborders, etc.

110. **(D)**   Adaptive radiation is a pattern that occurs when a lineage (single line of descent) branches into two or more lineages, and these further branch out. This pattern can occur when a species is able to invade environments that have previously been occupied by other species. In this case, when the dinosaurs became extinct, mammals invaded their vacated ecological niches and quickly diversified to adapt to the living conditions of the niches. In addition to invading vacant ecological niches, species can undergo adaptive radiation when they partition existing environments.

111. **(C)**   112. **(B)**   113. **(A)**   114. **(D)**

Note, from the illustration, the synapsis (pairing and attraction) of homologous chromosomes. This phenomenon, along with their crossing over denotes <u>prophase </u>of meiosis one. The chromosomes align along the central plane of the spindle in metaphase. The homologs separate in ana-phase with cell cytokinesis and complete separation into two daughter cells occurring in telophase. Illustration E is a meiosis II stage.

115. **(C)**   Hydrolysis is a type of chemical digestion. Amino acids are the digested building blocks of proteins. Glucose is a subunit of carbohydrates. Water molecules are required to split chemical bonds in hydrolysis but are not produced in the process.

PRACTICE TEST 2

CLEP Biology

Also available at the REA Study Center (*www.rea.com/studycenter*)

This practice test is also offered online at the REA Study Center. All CLEP exams are computer-based, and our test is formatted to simulate test-day conditions. We recommend that you take the online version of the test to receive these added benefits:

- **Timed testing conditions** – helps you gauge how much time you can spend on each question
- **Automatic scoring** – find out how you did on the test, instantly
- **On-screen detailed explanations of answers** – gives you the correct answer and explains why the other answer choices are wrong
- **Diagnostic score reports** – pinpoint where you're strongest and where you need to focus your study

PRACTICE TEST 2

CLEP Biology

(Answer sheets appear in the back of the book.)

TIME: 90 Minutes
115 Questions

DIRECTIONS: Each of the questions or incomplete statements below is followed by five possible answers or completions. Select the BEST choice in each case and fill in the corresponding oval on the answer sheet.

1. Which of the following factors exerts the most influence over limiting cell size?

   (A) A rigid cell wall
   (B) The ratio of surface area to volume of cytoplasm
   (C) The replication of mitochondria
   (D) The chemical composition of the cytoplasm
   (E) The chemical composition of the cell membrane

2. When a hamburger is consumed by an individual, it passes through all the following organs EXCEPT the

   (A) mouth
   (B) esophagus
   (C) salivary gland
   (D) stomach
   (E) small intestine

3. Which of the following conducted early research into the process of genetic inheritance?

   (A) Hooke
   (B) Mendeleev
   (C) Schwann
   (D) Schleiden
   (E) Mendel

4. In order for a species to be established on an island, it must have all of the following features EXCEPT

   (A) a population large enough to ensure successful reproduction
   (B) a food source for the species
   (C) a predator of the species
   (D) a suitable habitat
   (E) a source of moisture

5. Which of the following are NOT involved in the immune system?

   (A) Antibodies
   (B) Stem cells
   (C) T cells
   (D) Epithelial cells
   (E) B cells

6. In what way may a mass extinction event allow for diversification of species?

   (A) Mass extinction opens up ecological niches, making conditions favorable for the establishment of new, diverse species.
   (B) Mass extinction encourages convergent evolution.
   (C) Mass extinction serves to magnify evolution of altruistic traits.
   (D) Mass extinction allows for preservation of genetic material through fossilization.
   (E) Mass extinction occurs due to extreme climate changes.

7. Proteins may contain all of the following elements EXCEPT

   (A) carbon
   (B) hydrogen
   (C) magnesium
   (D) oxygen
   (E) nitrogen

8. Which of the following are within the phylum Chordata?

   (A) Crabs
   (B) Flat worms
   (C) Nematodes
   (D) Mollusks
   (E) Snakes

9. The process that releases energy for use by the cell is known as

   (A) photosynthesis
   (B) aerobic metabolism
   (C) anaerobic metabolism
   (D) cellular respiration
   (E) anabolism

10. The weakest type of chemical bonds of those listed are

   (A) ionic bonds
   (B) hydrogen bonds
   (C) double bonds
   (D) disulfide bonds
   (E) covalent bonds

11. All of the following are myths or misconceptions about the evolution of *Homo sapiens* EXCEPT:

   (A) *Homo sapiens* evolved from chimpanzees.
   (B) There is a linear sequence, or ladder, of different primates that leads to *Homo sapiens*.
   (C) The large brain and upright posture of *Homo sapiens* evolved together.
   (D) Chimpanzees are more closely related to *Homo sapiens* than to other apes.
   (E) *Homo sapiens* evolved in North America.

12. The stomach secretes all of the following EXCEPT

   (A) digestive enzymes
   (B) hydrochloric acid
   (C) gastric juices
   (D) acetic acid
   (E) mucous

13. Algae and protozoa are organisms within which kingdom?

   (A) Plantae
   (B) Animalia
   (C) Mammalia
   (D) Fungi
   (E) Protista

14. Which of the following shows the effect of substrate concentration on the initial reaction rate in the presence of a limited amount of enzyme?

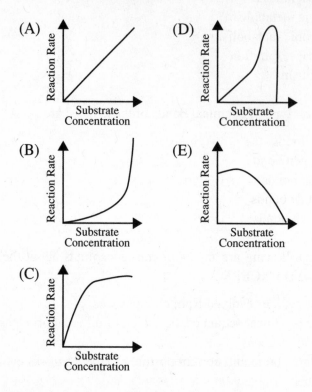

15. Plants and animals obtain usable nitrogen through the action of

(A) respiration
(B) nitrogen fixing by bacteria and lightning
(C) nitrogen processing in the atmosphere
(D) digestion
(E) catabolism

16. When the water pressure is equal inside and outside the cell, it is said to be

(A) hydrostatic
(B) diffuse
(C) isohydric
(D) isotonic
(E) hydrophobic

17. Energy transformations that occur as chemicals are broken apart or synthe-sized within the cell are collectively known as

    (A) catabolism
    (B) anabolism
    (C) metabolism
    (D) synthesis
    (E) hydrolysis

18. All of the following statements about photosynthesis are true EXCEPT:

    (A) Photosynthesis occurs through numerous small steps.
    (B) Photosynthesis can be summarized by the equation:

    $$6CO_2 + 6H_2O + \text{light energy} \rightarrow C_6H_{12}O_6 + 6O.$$

    (C) Chlorophyll is not harmed or used up by the photosynthetic process.
    (D) Photosynthesis occurs in green plants and algae.
    (E) Chlorophyll speeds the photosynthetic process, but it is not required for photosynthesis to occur.

19. The largest number of ATP molecules is formed by

    (A) fermentation
    (B) the electron transfer system
    (C) glycolysis
    (D) the Krebs cycle
    (E) photolysis

Questions 20–24 consist of five lettered terms followed by a list of numbered phrases. For each numbered phrase select the one term that is most closely re-lated to it. Each term may be used once, more than once, or not at all.

    (A) Differential reproduction
    (B) Mutation
    (C) Gene migration
    (D) Hardy-Weinberg Equilibrium
    (E) Allopatric speciation

20. Accidental change in genetic material

21. Best-adapted individuals are most likely individuals to reproduce viable offspring.

22. In a population that is in equilibrium with its environment, gene frequencies and genotype ratios remain constant.

23. The evolution of two separate species due to two populations being geographically isolated from each other

24. The infiltration of a previously isolated group by a member of an adjacent population, resulting in the increase of genetic possibilities

25. How does DNA produce particular genetic traits?

   (A) Through DNA replication
   (B) Through protein synthesis
   (C) Through genetic imprinting
   (D) Through genetic maintenance
   (E) Through genetic screening

26. A single DNA strand that has the sequence GATACCA would be complemented by a strand of DNA with which sequence?

   (A) GATACCA
   (B) CTATGGT
   (C) ACCATAG
   (D) CUAUCCU
   (E) UCCUAUC

27. The longest of the two major periods of the cell cycle in which the cell is carrying on its primary function is known as

   (A) interphase
   (B) prophase
   (C) telophase
   (D) anaphase
   (E) metaphase

Questions 28–32 consist of five lettered terms followed by a list of numbered phrases. For each numbered phrase, select the one term that is most closely related to it. Each term may be used once, more than once, or not at all.

(A) Centriole
(B) Free ribosomes
(C) Lysosomes
(D) Endocytic vesicle
(E) Nucleus

28. Digestive organelles that also serve to break up dead cell material

29. Site of protein synthesis

30. Membrane sack that transports large molecules through the cell membrane

31. Site of chromosomal replication

32. Pinwheel arrangement of microtubules that form structural skeleton

Questions 33–37

Normal skin color in mice is dominant to albino. In the following questions, N stands for normal skin color and n for albino.

33. Three offspring of two normal skinned parents have normal skin, but one is albino. Which of the following must be true?

(A) One parent must have the NN genotype.
(B) Both parents must have the NN genotype.
(C) One parent must have the nn genotype.
(D) Both parents must have the Nn genotype.
(E) Neither parent has the Nn genotype.

34. If these two normal skinned parents have eight offspring, approximately how many are likely to be albino?

(A) 2
(B) 4
(C) 1
(D) 8
(E) 3

35. The albino offspring from the F_1 generation described above produces one albino offspring and one normal offspring in the F_2 generation. What must be the genotype of the albino's mate?

    (A) Either Nn or NN
    (B) Either Nn or nn
    (C) Nn
    (D) nn
    (E) Both Nn and nn

36. What percentage of the offspring of two albino parents would most likely be normal?

    (A) 100%
    (B) 50%
    (C) 25%
    (D) 10%
    (E) 0%

37. What are the chances that two normal parents each carrying recessive genes for albinism could have a heterozygous normal offspring?

    (A) 1 out of 2
    (B) 3 out of 4
    (C) 2 out of 3
    (D) 0 out of 4
    (E) 2 out of 2

38. Which of the following is NOT an energy storage molecule?

    (A) Cellulose
    (B) Ribonucleic acid
    (C) Starch
    (D) Lipid
    (E) Sugar

39. The organelle of a cell that engages in both passive and active transport is the

    (A) rough endoplasmic reticulum
    (B) smooth endoplasmic reticulum
    (C) Golgi complex
    (D) centriole
    (E) cell (plasma) membrane

Questions 40–42 consist of five lettered terms followed by a list of numbered phrases. For each numbered phrase, select the one term that is most closely related to it. Each term may be used once, more than once, or not at all.

    (A) Aggregate fruit
    (B) Simple fruit
    (C) Seed
    (D) Monocot
    (E) Multiple fruit

40. Fused ovaries of a single flower (e.g., raspberry)

41. Several ovaries of separate flowers (e.g., pineapple)

42. Single ripened ovary (e.g., apple)

43. Stem tissue includes all of the following EXCEPT

    (A) vascular tissue
    (B) xylem
    (C) phloem
    (D) cuticle
    (E) sieve plates

44. An ion that binds to an enzyme making it more able to catalyze a reaction is known as

    (A) a protein
    (B) an inorganic cofactor
    (C) a coenzyme
    (D) a prosthetic group
    (E) an inhibitor

45. Which type of tissue is made up of stacked cells connected by sieve plates that allow nutrients to pass from cell to cell?

    (A) Xylem
    (B) Meristem
    (C) Phloem
    (D) Internodal
    (E) Ectoderm

46. All of the following characteristics of water make it valuable to living organisms EXCEPT

    (A) transparency
    (B) polarity
    (C) lower density when solid than when liquid
    (D) high specific heat
    (E) pH of 11

47. Each ecosystem can support a certain number of organisms; this number is known as the

    (A) natality
    (B) population
    (C) carrying capacity
    (D) community
    (E) biosphere

48. Many insects have special respiratory organs known as _____.

    (A) spiracles
    (B) alveoli
    (C) the cephalothorax
    (D) lungs
    (E) lymphocytes

49. What happens to most chemical pollutants that are accidentally ingested by a human?

   (A) They are broken down, mixed with broken down pigment molecules in the bile, and excreted in the feces.
   (B) They are attacked by the lymphatic system and recycled in protein synthesis.
   (C) They are collected in lymph tissue and excreted through the skin.
   (D) They build up in epithelial tissue until lethal levels are reached.
   (E) They are locked within the mitochondria.

50. The process of forming egg and sperm cells in the reproductive organs is known as

   (A) gametogenesis
   (B) spermatagonium
   (C) oogonium
   (D) gametocide
   (E) mitosis

51. The cells of a developing embryo (at the gastrula stage) differentiate into layers (called germ layers) that will later develop into various tissues and organs. Which layer will eventually form muscles, skeletal organs, and the circulatory, respiratory, reproductive, and excretory systems?

   (A) Endoderm
   (B) Ectoderm
   (C) Blastula
   (D) Mesoderm
   (E) Morula

Questions 52–56 consist of five lettered terms followed by a list of numbered phrases. For each numbered phrase, select the one term that is most closely related to it. Each term may be used once, more than once, or not at all.

(A) Circadian rhythm
(B) Fixed action pattern
(C) Altruism
(D) Imprinting
(E) Habituation

52. Cycle of daily behavior based on an internal clock and environmental clues

53. Behavior that benefits the group at the individual's expense

54. Innate behavior that is independent of the environment

55. Learned behavior that results in not responding to a stimulus

56. Behavior learned only during a critical period of an organism's life

57. The physical place where a particular organism lives is called its

(A) niche
(B) biosphere
(C) lithosphere
(D) habitat
(E) hydrosphere

58. A _____ is a length of DNA (with corresponding histones) that is responsible for the production of a particular protein that causes a particular trait to be expressed in an organism.

(A) chromosome
(B) mutation
(C) genome
(D) gene
(E) nucleotide

59. All of the following are requirements of the habitat of an apple tree EXCEPT

   (A) soil quality
   (B) available sunlight
   (C) seasonal temperature fluctuations
   (D) amount of rainfall
   (E) role in the food chain

60. Sugars synthesized by photosynthesis travel through _____ _____ to various parts of the plant.

   (A) epidermal tissue
   (B) vascular bundles
   (C) meristem tissue
   (D) parenchyma tissue
   (E) cuticle tissue

61. Which of the following environmental factors is NOT recycled?

   (A) Nitrogen
   (B) Carbon
   (C) Phosphorous
   (D) Water
   (E) Silicon

62. As energy is transferred through the trophic levels, some energy

   (A) is created by producers
   (B) is destroyed by producers
   (C) is destroyed by decomposers
   (D) becomes unusable
   (E) is created by autotrophs

63. Channels in cell membranes that carry water between cells are called

   (A) guard cells
   (B) plasmodesmata
   (C) stomata
   (D) internodes
   (E) nodes

64. All of the following are steps in the phosphorous cycle EXCEPT:

    (A) Phosphorous becomes available for erosion as undersea sedimentary rocks are up-thrust by volcanic activity.
    (B) Phosphorous is returned to the ground through animal waste.
    (C) Gaseous phosphorous is absorbed from the atmosphere by plant leaves.
    (D) Erosion releases phosphorous from rocks into streams where it combines with oxygen to form phosphates in lakes that are then absorbed by plants.
    (E) Phosphorous is recycled through the food chain as animals consume plants and other animals.

65. A bilayer of phospholipids with protein globules interspersed is characteristic of which of the following organelles?

    (A) Cell membrane
    (B) Mitochondria
    (C) Lysosome
    (D) Chromatin
    (E) Nucleus

66. A pond ecosystem has sharp boundaries at the shorelines. The sharp boundary of an ecosystem is known as

    (A) a segregation point
    (B) a succession
    (C) an ecotone
    (D) a borderline
    (E) a dispersion region

Questions 67–71 consist of five lettered terms followed by a list of numbered phrases. For each numbered phrase, select the one term that is most closely related to it. Each term may be used once, more than once, or not at all.

    (A) Cenozoic
    (B) Pangea
    (C) Paleozoic
    (D) Mesozoic
    (E) Precambrian

67. Began with the Cambrian explosion; representatives of most modern phyla present

68. Fossilized burrows found in rocks of this age indicate the development of multicellular animals but only with soft parts.

69. This era ended with the extinction of dinosaurs.

70. This era includes the radiation of angiosperms.

71. This era includes the so-called Age of Fishes because of the extensive radiation of fish.

72. Which of the following cell organelles is known as the cell's "powerhouse" because it produces energy for the cell's use?

    (A) Nucleus
    (B) Mitochondrion
    (C) Smooth endoplasmic reticulum
    (D) Ribosome
    (E) Cell membrane

73. Which one of the following statements about cells is NOT true?

    (A) All living things are made up of one or more cells.
    (B) Cells may be seen with a microscope.
    (C) Cells are the basic units of life.
    (D) All cells come from pre-existing cells.
    (E) All cells have cell walls.

Questions 74–78 consist of five lettered terms followed by a list of numbered phrases. For each numbered phrase, select the one term that is most closely related to it. Each term may be used once, more than once, or not at all.

    (A) Trachea
    (B) Pharynx
    (C) Bronchi
    (D) Larynx
    (E) Alveoli

74. Surrounded by capillaries that allow for carbon dioxide to diffuse into the lungs and oxygen to diffuse out

75. Includes the larynx and the glottis

76. Branched tubes that lead to lungs

77. Contains the vocal cords

78. Between the nasal passage and the trachea

79. The bee belongs to which phylum?

    (A) Arthropoda
    (B) Aves
    (C) Annelida
    (D) Nematoda
    (E) Porifera

80. Which of the following is a polymer of amino acids?

    (A) Lactose
    (B) Lactase
    (C) Glycogen
    (D) Sucrose
    (E) Cellulose

81. Vertebrates with no jaws belong to the super-class

    (A) Aganatha
    (B) Gnathostomata
    (C) Protista
    (D) Cnidaria
    (E) Porifera

82. The synthesis of ATP molecules to store energy is an example of

    (A) anabolism
    (B) catabolism
    (C) adaptive radiation
    (D) lysis
    (E) inhibition

83. Each of the following reactions may occur after glycolysis EXCEPT

   (A) photolysis
   (B) aerobic respiration
   (C) the Krebs cycle
   (D) the electron transport cycle
   (E) fermentation

84. What type of organic molecule has this group attached?

   (A) Carbohydrate
   (B) Lipid
   (C) Base
   (D) Organic acid
   (E) Glycerol

85. Which of the following structures provide rigidity to plant cells but not animal cells?

   (A) Microtubules
   (B) Cell walls
   (C) Microfilaments
   (D) Centrioles
   (E) Mitochondria

Questions 86–90 consist of five lettered terms followed by a list of numbered phrases. For each numbered phrase, select the one term that is most closely related to it. Each term may be used once, more than once, or not at all.

   (A) Nodes
   (B) Nonvascular plants
   (C) Angiosperms
   (D) Gymnosperms
   (E) Lateral buds

86. Bryophytes

87. Flowering plants

88. Monocots and dicots

89. Produce seeds without flowers

90. Conifers and cycads

91. Which of the following represents an opportunistic life strategy (r-selection)?

    (A) Lichens invade a bare rock area after a volcanic eruption.
    (B) Coniferous trees spread to an adjacent area.
    (C) Lightning wipes out a forest of deciduous trees.
    (D) A species of mice emigrate into a forest community.
    (E) Broad-leaf trees flourish during the rainy season.

92. The nervous system is an integrated circuit with many functions. Which of the following parts of the nervous system are matched with the wrong function?

    (A) Forebrain—controls olfactory lobes (smell)
    (B) Cerebrum—controls function of involuntary muscle
    (C) Hypothalamus—controls hunger and thirst
    (D) Cerebellum—controls balance and muscle coordination
    (E) Midbrain—contains optic lobes, controls sight

93. Which of the following statements about the cell theory is NOT true?

    (A) It was developed by the German scientists Schleiden and Schwann.
    (B) It states that all living things are made of cells.
    (C) It states that cells are the basic units of life.
    (D) It states that all cells come from pre-existing cells.
    (E) It states that anaerobic cells existed before aerobic cells.

94. Nitrogen is made available to organisms of the food chain through all of the following EXCEPT:

   (A) Bacteria break ammonia into nitrites, then into nitrates that are usable by plants.
   (B) Volcanic activity produces ammonia and nitrates that enter the soil and can be absorbed by plants.
   (C) Lightning reacts with atmospheric nitrogen to form nitrates that are absorbed by plants.
   (D) Nitrogen is recycled from dead organisms and reenters the food chain.
   (E) Nitrogen is absorbed into ocean water.

95. Which of the following statements about enzymes is NOT true?

   (A) High temperatures destroy most enzymes.
   (B) Enzymes only function within living things.
   (C) An enzyme is unaffected by the reactions it catalyzes, so it can be used over and over again.
   (D) Enzymes are usually very specific to certain reactions.
   (E) Some enzymes contain a non-protein component that is essential to their function.

96. Scurvy is a disease caused by a lack of vitamin C in which the body is unable to build enough collagen (a major component of connective tissue). The most plausible explanation for this malfunction is

   (A) vitamin C is an amino acid component of collagen
   (B) vitamin C is a coenzyme required in the synthesis of collagen
   (C) vitamin C destroys collagen
   (D) vitamin C is produced by collagen
   (E) vitamin C is another name for collagen

97. In ferns, the individual we generally recognize as an adult fern is really which structure?

   (A) A mature gametophyte
   (B) A prothallus
   (C) A mature sporophyte
   (D) A young sporophyte
   (E) A young gametophyte

Questions 98–100 consist of five lettered terms followed by a list of numbered phrases. For each numbered phrase, select the one term that is most closely related to it. Each term may be used once, more than once, or not at all.

   (A) Savanna
   (B) Tropical rain forest
   (C) Taiga
   (D) Desert
   (E) Tundra

98. Extreme cold, low precipitation, modified grassland, permafrost

99. Extreme heat or cold, sparse vegetation, very low precipitation, reptiles

100. Warm temperatures, moderate precipitation, grassland

101. When a stem bends toward the light, it is due to

   (A) the increased level of auxin on the light side of the shoot tip
   (B) the migration of auxin toward the dark side of the shoot tip
   (C) the migration of auxin toward the light side of the shoot tip
   (D) the elongation of cells on the light side of the shoot tip
   (E) the decreased level of auxin on the dark side of the root tip

102. The theory of punctuated equilibrium assumes that

      I.   There are periods of stability during which little evolutionary change occurs.
     II.   Speciation can occur rapidly over a very short period of time.
    III. Evolution occurs gradually within lineages.

   (A) I only
   (B) II only
   (C) III only
   (D) I and II
   (E) I, II, and III

103. Hemophilia is a disease caused by a sex-linked recessive gene on the X chromosome; therefore,

(A) females have twice the likelihood of having the disease, since they have two X chromosomes
(B) mothers can pass the gene with probability to either a son or daughter
(C) females can never have the disease, but can only be carriers
(D) inbreeding has no effect on the incidence of the disease, since it is purely sex-linked
(E) a hemophiliac son is always produced if his father has the gene and hence, the disease

104. Because fungi can obtain nutrients from nonliving organic matter, they are referred to as

(A) parasitic
(B) saprophytic
(C) eukaryotic
(D) heterotrophic
(E) pathogenic

Questions 105–108 refer to the diagram below.

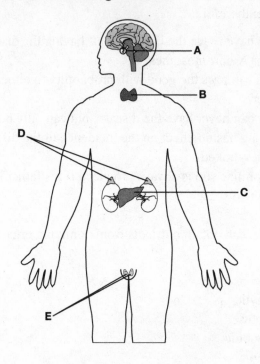

105. A sudden change in the amount of extracellular fluid will be corrected by events following release of substances from this organ.

106. A person eats three candy bars. Within minutes, this endocrine gland affects blood-glucose homeostasis.

107. Substances that cause vasoconstriction change the diameters of blood vessels in order to assist in increasing blood pressure. Such substances are produced by this gland.

108. Secrete aldosterone

109. Restriction enzymes are used in genetic research to

    (A) cleave DNA molecules at certain sites
    (B) produce individual nucleotides from DNA
    (C) slow down the reproductive rate of bacteria
    (D) remove DNA strands from the nucleus
    (E) prevent histones from reassociating with DNA

110. Select the most abundant element of protoplasm from the following choices:

(A) calcium
(B) carbon
(C) phosphorus
(D) sulfur
(E) zinc

111. All of the following fates for sugar produced during photosynthesis are possible in a plant cell, EXCEPT

(A) its polymerization into starch for storage purposes
(B) its decomposition for energy production
(C) its polymerization into glycogen for storage purposes
(D) its use in the synthesis of other organic molecules
(E) its use in the synthesis of sucrose

112. PKU (phenylketonuria) is an example of an inborn error of metabolism. These "errors" refer to

(A) congenital birth defects
(B) hormonal overproduction
(C) inherited lack of an enzyme
(D) nondisjunction
(E) atrophy of endocrine glands

113. All living organisms are classified as eukaryotes (true nucleus) or prokaryotes (before the nucleus). The only example of a prokaryote listed below is

(A) AIDS virus
(B) *E. coli*
(C) *Homo sapiens*
(D) an oak tree
(E) amoeba

114. Lichen, in which an alga and a fungus live in harmony, is an example of

    (A) mutualism
    (B) commensalism
    (C) parasitism
    (D) predation
    (E) competition

115. Darwin's theory of natural selection includes all of the following stipulations EXCEPT:

    (A) Every organism produces more organisms that can survive.
    (B) Due to competition, not all organisms survive.
    (C) Some organisms are more fit, i.e., they are able to survive better in the environment.
    (D) The difference in survivability is due to variations between organisms.
    (E) Variation is due, at least in part, to mutations.

PRACTICE TEST 2

Answer Key

1.	(B)	39.	(E)	77.	(D)
2.	(C)	40.	(A)	78.	(B)
3.	(E)	41.	(E)	79.	(A)
4.	(C)	42.	(B)	80.	(B)
5.	(D)	43.	(D)	81.	(A)
6.	(A)	44.	(D)	82.	(A)
7.	(C)	45.	(C)	83.	(A)
8.	(E)	46.	(E)	84.	(D)
9.	(D)	47.	(C)	85.	(B)
10.	(B)	48.	(A)	86.	(B)
11.	(D)	49.	(A)	87.	(C)
12.	(D)	50.	(A)	88.	(C)
13.	(E)	51.	(D)	89.	(D)
14.	(C)	52.	(A)	90.	(D)
15.	(B)	53.	(C)	91.	(A)
16.	(D)	54.	(B)	92.	(B)
17.	(C)	55.	(E)	93.	(E)
18.	(E)	56.	(D)	94.	(E)
19.	(B)	57.	(D)	95.	(B)
20.	(B)	58.	(D)	96.	(B)
21.	(A)	59.	(E)	97.	(C)
22.	(D)	60.	(B)	98.	(E)
23.	(E)	61.	(E)	99.	(D)
24.	(C)	62.	(D)	100.	(A)
25.	(B)	63.	(B)	101.	(B)
26.	(B)	64.	(C)	102.	(D)
27.	(A)	65.	(A)	103.	(B)
28.	(C)	66.	(C)	104.	(B)
29.	(B)	67.	(C)	105.	(A)
30.	(D)	68.	(E)	106.	(C)
31.	(E)	69.	(D)	107.	(D)
32.	(A)	70.	(A)	108.	(D)
33.	(D)	71.	(C)	109.	(A)
34.	(A)	72.	(B)	110.	(B)
35.	(C)	73.	(E)	111.	(C)
36.	(E)	74.	(E)	112.	(C)
37.	(A)	75.	(A)	113.	(B)
38.	(B)	76.	(C)	114.	(A)
				115.	(E)

PRACTICE TEST 2

Detailed Explanations of Answers

1. **(B)** The size of a cell is limited by the ratio of its surface area to volume. A cell will only remain stable if the surface area of the plasma membrane maintains a balance with the volume of cytoplasm. Only plant cells (and some bacteria) have cell walls. The replication of mitochondria, the chemical composition of the cytoplasm, and the chemical composition of the cell membrane do not specifically limit cell size.

2. **(C)** Ingested food does not pass through the salivary gland; rather, the saliva secreted from this gland enters the digestive tract and helps digest the food. Food does pass through the mouth, esophagus, stomach, and small intestine.

3. **(E)** Gregor Mendel studied the relationships between traits expressed in parents and offspring and the genes that caused the traits to be expressed. Hooke, Mendeleev, Schwann, and Schleiden were all scientists not directly involved with genetics.

4. **(C)** In order to become an established part of the island ecosystem a species must find a suitable ecological niche. Ultimately the species must be able to reproduce in its new setting, or it will not remain a part of the ecosystem. Organisms do not require a predator, though it may be necessary in some cases for population control. Many thriving organisms, including those that are at the highest trophic level, have no predators.

5. **(D)** Epithelial (skin) cells have no direct function in immunity. Cells involved in immunity are called lymphocytes and are produced in bone marrow as stem cells. B and T cells are two classes of lymphocytes, B cells and T cells. B cells produce antibodies into the bloodstream that find and attach themselves to foreign antigens (toxins, bacteria, etc.) and then both are destroyed. Some T cells patrol the blood for antigens, but T cells are also equipped to destroy antigens themselves. T cells also regulate immune responses.

6. **(A)** Extinction events cause ecological niches to be available for other, newer species to fill. None of the other answer choices is a true statement.

7. **(C)** Magnesium is not found within proteins. Carbon, hydrogen, oxygen, and nitrogen are all common elements found in proteins.

8. **(E)** Chordata have four defining characteristics: a notochord, a dorsal hollow nerve cord, pharyngeal gill slits, and a postanal tail during some point of their development. Of the given choices, only snakes have these characteristics during some point of their development.

9. **(D)** Cellular respiration is the process that releases energy for use by the cell. There are several steps involved in cellular respiration; some require oxygen (aerobic) and some do not (anaerobic). Photosynthesis is a type of anabolism, a reaction that harnesses and stores solar energy in chemical bonds.

10. **(B)** A hydrogen bond involves the attraction of atoms of different polarity and can be easily broken. Ionic bonds (where electrons are transferred) and covalent bonds (where electrons are shared), as well as double bonds and disulfide bridges, are all stronger than hydrogen bonds.

11. **(D)** Chimpanzees are more closely related to *Homo sapiens* than to other apes, but *Homo sapiens* did not evolve from chimpanzees. About five million years ago, the lineage that led to the modern *Homo sapiens* diverged from the lineage that led to the modern chimpanzee. It is a common misconception that the evolution leading to *Homo sapiens* occurred like a series of steps on a ladder. It is more like a branching tree with dead ends and new branches appearing simultaneously. The large brain and upright posture are important features of *Homo sapiens*, but the early hominoids stood upright before there was an increase in brain size. Most of the fossils of hominoids are from continents other than North America.

12. **(D)** The stomach secretes digestive enzymes, hydrochloric acid, and gastric juices, which all aid in digestion. The stomach also secretes mucous, which protects the stomach lining from the acids and gastric juices. The stomach does not secrete acetic acid.

13. **(E)** Algae and protozoa are within the Kingdom Protista, which contains one-celled eukaryotes. Kingdom Animalia contains organisms that are multicellular eukaryotes (including vertebrates and invertebrates). Mammalia is not a kingdom; it is a class within the subphylum vertebrata. Kingdom Fungi contains organisms that are multicellular eukaryotes, including molds and mushrooms. Kingdom Plantae contains organisms that are

multicellular, photosynthetic eukaryotes (including gymnosperms and angiosperms).

14. **(C)** An enzyme is a special protein that acts as a catalyst for organic reactions. A catalyst is a substance that changes the speed of a reaction without being affected itself. The enzyme will speed up the initial reaction rate as you increase the concentration of the substrate until all the enzymes are in use. Then the reaction rate will level off, shown by the graph—C.

15. **(B)** Neither plants nor animals are able to use nitrogen directly from the air. Instead, a process known as nitrogen fixing makes nitrogen available for absorption by the roots of plants. Nitrogen fixing is the process of combining it with either hydrogen or oxygen. Nitrogen fixing is accomplished in one of two ways—either by nitrogen-fixing bacteria or by the action of lightning. Nitrogen is present in the atmosphere (including the air we breathe), but it is not used in the process of respiration. Neither catabolism nor digestion renders nitrogen useful to living things.

16. **(D)** When the water concentration inside and outside the cell is equal, it is said to be in an isotonic state. Water will pass through the cell membrane by osmosis from an area of higher concentration to an area of lower concentration in order to produce isotonic conditions.

17. **(C)** Cellular metabolism is a general term that includes all types of energy transformation processes, including photosynthesis, respiration, growth, movement, etc. Energy transformations occur as chemicals are broken apart (catabolism) or synthesized within the cell (anabolism). Hydrolysis is a reaction that adds the elements of water (2 hydrogen, 1 oxygen) to another compound.

18. **(E)** Though the process of photosynthesis actually occurs through numerous small steps, the entire process can be summed up with the following equation:

$$6CO_2 + 6H_2O + \text{light energy} \rightarrow C_6H_{12}O_6 + 6O_2$$

Chlorophyll is a green pigment that must be present in order for photosynthesis to occur. Chlorophyll has the ability of absorbing a photon of light and is found in the grana of the chloroplast. Chlorophyll is not used up in the photosynthetic process.

19. **(B)** The electron transfer system (ETS) produces the most ATP molecules, yielding 34 ATPs per glucose molecule. Fermentation and glycolysis each produce two ATPs per glucose molecule. Some of the products of the Krebs cycle are easily converted to ATP, but the main energy products of the Krebs cycle are those that liberate electrons that are then used in the electron transfer reactions. Photolysis is a reaction of photosynthesis. In photolysis, chlorophyll pigments absorb photons of light, leaving the chlorophyll in an excited (higher energy) state. The excited chlorophyll then supplies energy to a series of reactions that produce ATP from ADP and inorganic phosphate (P_i).

20. **(B)** A mutation is an accidental change of the DNA sequence of a gene that can result in creating a change of trait that is not found in the parent.

21. **(A)** Differential reproduction proposes that those individuals within a population that are most adapted to the environment are also the most likely individuals to reproduce viable offspring.

22. **(D)** The Hardy-Weinberg Law of Equilibrium states that where random mating is occurring within a population that is in equilibrium with its environment, the gene frequencies and genotype ratios will remain constant from generation to generation. The Hardy-Weinberg Equation is a mathematical formula that shows why recessive genes do not disappear over time from a population.

23. **(E)** Allopatric speciation occurs when two populations are geographically isolated from each other. Over time this results in the production of two separate species.

24. **(C)** When an individual from an adjacent population of the same species immigrates and breeds with a member of a previously locally isolated group, it results in a change of the gene pool, known as gene migration.

25. **(B)** The primary role of DNA in the cell is the control of protein synthesis. Genetic traits are expressed and specialization of cells occurs as a result of the combination of proteins produced by the DNA of a cell. DNA replication allows for the genetic code to be preserved in future generations of cells. When expression of genetic traits is determined by whether the trait is inherited from the mother or the father, it is called genetic imprinting. Genetic maintenance simply refers to the preservation of the integrity of genetic information from one generation to another. Genetic screening is the systematic search for individuals with a specific genotype in a delineated population.

26. **(B)** In DNA, guanine (G) pairs with cytosine (C) and thymine (T) with adenine (A). Therefore, the sequence GATACCA would pair with the sequence CTATGGT.

27. **(A)** There are two major periods within the cell cycle: interphase and mitosis. Interphase is the period when the cell is active in carrying on the function it was designed to perform within the organism. Cells spend much more time in interphase than in the cell division portion of their cycle, which includes the prophase, telophase, anaphase, and metaphase phases of mitosis plus cytokinesis.

28. **(C)** Lysosomes are membrane-bound organelles that contain digestive enzymes that digest dead or unused material within the cell or materials absorbed by the cell for use.

29. **(B)** Protein synthesis occurs at the free ribosomes. Free ribosomes float unattached within the cytoplasm. They contain RNA that is specific to their function (ribosomal RNA or rRNA) in protein formation.

30. **(D)** Endocytic vesicles are formed when the plasma membrane of a cell encloses a molecule outside the membrane, then releases a membrane-bound sack containing the desired molecule into the cytoplasm. This process allows the cell to absorb molecules that are larger in size than would be able to pass through the cell membrane.

31. **(E)** The nucleus contains the chromosomes and is the site of reproduction through mitosis and meiosis.

32. **(A)** Centrioles are tubes constructed of a geometrical arrangement of microtubules in a pinwheel shape. Their function includes the formation of new microtubules, but is primarily to form the structural skeleton around which cells split during mitosis and meiosis.

33. **(D)** As shown in the following Punnett square, the only way offspring could be albino is if each parent has at least one recessive gene for albinism. Since both have a normal phenotype, both must have the Nn genotype. Also, each parent must also have one recessive allele for albinism in order to produce one nn child.

	N	n
N	NN	Nn
n	Nn	nn

34. **(A)** The phenotype ratio for this generation of offspring will be 3 normal: 1 albino; so out of 8 offspring, 2 are likely to be albino.

35. **(C)** The albino parent from the F_1 cross must have the nn genotype. As shown in the following Punnett square, if the mate was albino (nn), all the offspring would be albino; and if the mate was homozygous (NN), then all the offspring would be normal. In order to produce both phenotypes, the second parent must be heterozygous (Nn).

	n	n
N	Nn	Nn
n	nn	nn

36. **(E)** All offspring of two albino parents (each must have the genotype nn) will be albino, so the answer is 0%.

37. **(A)** Referring to the Punnett square in answer No. 33, it is clear that the Nn × Nn cross would yield 1 out of 4 albino (homozygous), 1 out of 4 homozygous normal (NN), and 2 out of 4 heterozygous normal (Nn) children.

38. **(B)** Ribonucleic acid (RNA) is a molecule that stores information for protein synthesis and genetic coding, not energy. Cellulose, starch, lipid, and sugar molecules all store energy within their chemical bonds.

39. **(E)** The selective permeability of the cell membrane serves to manage the concentration of substances within the cell, preserving its health. There are two methods by which substances can cross the cell membrane, passive transport and active transport. The rough endoplasmic reticulum and smooth endoplasmic reticulum serve as channels for moving molecules through a cell. The Golgi complex stores and packages proteins and lipids for transport and use throughout the cell. Centrioles provide support to the cell, especially during replication.

40. **(A)** An aggregate fruit is a compound fruit that develops from many ovaries of a single flower fusing together (e.g., raspberry).

41. **(E)** A multiple fruit is a compound fruit that forms from several ovaries of separate flowers that fuse together during ripening (e.g., strawberry or pineapple).

42. **(B)** Fruits that are developed from a single ripened ovary are known as simple fruits (e.g., apple, olive, acorn, cucumber).

43. **(D)** Stem tissue does not include the cuticle. The cuticle is found covering the leaf. The stem is made of vascular tissue, including both xylem and phloem. Sieve plates exist between cells of the stem.

44. **(D)** Prosthetic groups, which may be ions or non-protein molecules, are similar to cofactors in that they facilitate the enzyme reaction. However, prosthetic groups are tightly attached by covalent bonds to the enzyme, rather than being separate atoms or molecules. The enzyme itself is a protein; the prosthetic group is not. An inorganic cofactor does not bind with the enzyme. A coenzyme also does not bind to the enzyme itself and is not an ion. An inhibitor attaches to the enzyme, but it blocks the enzyme reaction rather than enhancing it.

45. **(C)** Phloem tissue, made of stacked cells connected by sieve plates (that allow nutrients to pass from cell to cell), transports food made in the leaves (by photosynthesis) to the rest of the plant. Xylem tissue transfers water and does not require sieve plates to allow nutrients through. Meristem is the tissue that is found in the root cap and is responsible for quick growth in the roots. Internodal tissue is found on the stem between nodes. Ectoderm is the outermost of the three main layers of an embryo.

46. **(E)** Water has a pH of approximately 7, making it neither basic (under 7) nor alkaline (over 7). Transparency, polarity, density, and specific heat are characteristics that make water valuable to living things.

47. **(C)** Carrying capacity is the number of organisms that can be supported within a particular ecosystem. The term *natality* refers to the birthrate of a population. The population includes the number of organisms in a given community, whether or not the community is at its carrying capacity. The community is comprised of all the organisms that interact within a given ecosystem whether or not it is at carrying capacity or not. The biosphere includes all the living and nonliving components of the Earth that allow the Earth to support living things.

48. **(A)** Insects use spiracles for gas exchange. Alveoli are found within the lungs (lungs are not found in insects). A cephalothorax is the head and thorax of arachnids and crustaceans, and is not found in insects. Lymphocytes are a type of cell found in blood and lymph tissue.

49. **(A)** The liver filters out most chemical pollutants, which are then mixed with broken-down pigments in the bile. Bile is secreted into the small intestine, then proceeds to the large intestine and is expelled in the feces.

50. **(A)** Egg and sperm cells are called gametes and are formed in the process of gametogenesis. Spermatagonium are cells that eventually may become sperm cells, and oogonium are cells that eventually may become egg cells. Gametocide refers to the destruction of gametes. Mitosis is the reproduction of a parent cell to produce identical daughter cells.

51. **(D)** The mesoderm (between the ectoderm and endoderm) layer will eventually form muscles and organs of the skeletal, circulatory, respiratory, reproductive, and excretory systems. The endoderm will become the gut lining and some accessory structures. The ectoderm will become the skin, some endocrine glands, and the nervous system. The morula is the cluster of cells that results from cleavage of the original cells. The blastula develops from the morula as a thin layer of cells surrounding an internal cavity.

52. **(A)** Circadian rhythms consist of an organism's daily repeated behavior such as wake and sleep cycles that function according to its internal clock. The internal clock is affected by environmental cues such as hours of sunlight per day, etc.

53. **(C)** Altruism is a social behavior of an organism that is beneficial to the group at the individual's expense.

54. **(B)** A fixed action pattern (FAP) is a type of innate behavior (instinct). The FAP is a pre-programmed response to a particular stimulus (known as a releaser or a sign stimulus). FAPs include courtship behaviors and feeding of young. Organisms automatically perform FAPs without any prior experience (FAPs are not learned).

55. **(E)** Habituation occurs when an individual learns not to respond to a particular stimulus, for instance, when a stimulus is repeated many times without consequence.

56. **(D)** Imprinting is behavior that is learned during a critical point (often very early) in an individual's life. For instance, a gosling is imprinted with the impression of its mother immediately after hatching. Imprinting enables the young to recognize members of their own species, etc.

57. **(D)** A habitat refers to the physical place where an organism lives. A species' habitat must include all the factors that will support its life and reproduction. The niche also includes the role played by the organism in its food chain. The biosphere is the part of the Earth that contains all living things. The lithosphere (Earth) and hydrosphere (water) are parts of the biosphere.

58. **(D)** A gene is the portion of DNA that produces a particular expressed trait. A chromosome contains many genes and is a structure comprised of linear DNA and associated proteins. A mutation is a mistake in DNA replication. The genome is the total amount of genetic information available for a given species. Nucleotides are the monomers (containing a sugar, a phosphate group, and a nitrogenous base) that form nucleic acids.

59. **(E)** A species' role in the food chain is a part of its niche, not its habitat. The niche includes the habitat, but the habitat is within the niche. A species' habitat includes all the factors that will support its life and reproduction. These factors may be biotic (population, food source, etc.) and abiotic (i.e., nonliving—weather, temperature, soil features, sunlight, etc.).

60. **(B)** The sugars produced by photosynthesis are transported throughout the plant via the vascular bundles. The vascular bundles make up the veins in the leaf and are also distributed throughout the stem. Epidermal tissue is the outermost layer of cells of the stem. Meristem tissue consists of undifferentiated cells capable of quick growth and specialization. Meristem tissue is responsible for elongation of the stem. Parenchyma tissue has loosely packed cells that allow for gas and moisture exchange. The cuticle is the waxy outer coating of leaves.

61. **(E)** Carbon, nitrogen, phosphorous, and water are all recycled through biogeochemical cycles. Silicon is the major component of sand and is the most abundant element found in the lithosphere. It is not recycled through environmental processes.

62. **(D)** The energy cycle of the food chain is subject to the laws of thermodynamics—no energy can be created or destroyed, and as energy changes form and passes from one level to another, some becomes unusable.

63. **(B)** One pathway for water to pass through cell walls and plasma membranes toward xylem tissue is through an intercellular route through channels in the cell membranes known as plasmodesmata. Guard cells and stomata are whole cells that regulate the intake and outflow of water, whereas plasmodesmata are channels within a cell membrane. Nodes are places on a stem where a leaf can begin to grow, and internodes are the areas on the stem between nodes.

64. **(C)** Phosphorous is nearly always found in solid form, within rocks and soil. Phosphorous gas is very rare and is not absorbed by plant leaves. The usable reservoir of Earth's phosphorous is found within rocks and soil. Water dissolves phosphorous from rocks by erosion and carries it into

rivers and streams. Here phosphorous and oxygen unite to form phosphates that end up in bodies of water. Phosphates are absorbed by plants in and near the water and are used in the synthesis of organic molecules. As in the carbon and nitrogen cycles, phosphorous is then passed through the food chain and returned through animal wastes and organic decay of dead matter. New phosphorous enters the cycle as undersea sedimentary rocks are upthrust during the shifting of the Earth's tectonic plates. New rock-containing phosphorous is then exposed to erosion and enters the cycling process.

65. **(A)** The cell membrane is composed of a double layer (bilayer) of phospholipids with protein globules imbedded within the layers. The construction of the membrane allows it to aid cell function by permitting entrance and exit of molecules as needed by the cell. Mitochondria are the organelles in which cellular respiration occurs. A lysosome is a packet of digestive enzymes that destroy cellular wastes. Chromatin is disorganized DNA with histones attached. The nucleus is the organelle in which cellular reproductive processes occur.

66. **(C)** The sharp boundary of a community is called an ecotone. None of the other terms applies to the boundary of an ecological community.

67. **(C)** The Cambrian period is the earliest period of the Paleozoic era. The Cambrian explosion of life resulted in representatives of most of the modern phyla being present.

68. **(E)** Fossilized burrows from multicellular organisms begin to appear in the geological record approximately 700 million years ago during the Precambrian period. These multicellular animals had only soft parts that could not be fossilized.

69. **(D)** The end of the Mesozoic Era is defined by the extinction of dinosaurs.

70. **(A)** The Cenozoic is the most recent and present era. It includes the radiation of flowering plants, the angiosperms.

71. **(C)** There was extensive radiation of fish during the Devonian and Silurian periods within the Paleozoic Era.

72. **(B)** Mitochondria are called the cell's "powerhouses," as they constitute the center of cellular respiration. Cellular respiration is the process of breaking up covalent bonds within sugar molecules with the intake of oxygen and release of energy in the form of ATP [adenosine tri-phosphate]

molecules. ATP is the energy form used by all cell processes. Mitochondria (singular of mitochondrion) are found wherever energy is needed within the cell, and are more numerous in cells that require more energy (muscle, etc.). The nucleus contains mitochondria, but mitochondria also exist outside the nucleus. The smooth endoplasmic reticulum is a system of channels for moving substances within the cell. Ribosomes are the site of protein synthesis within the cell. The cell membrane encloses the cell and allows for substances to pass in and out of the cytoplasm.

73. **(E)** Only plant cells have cell walls, but all cells have a cell/plasma membrane.

74. **(E)** At the alveoli, carbon dioxide is diffused into the lungs and oxygen replaces the carbon dioxide in the blood.

75. **(A)** The trachea includes the windpipe or larynx in its upper portion and the glottis, an opening that allows the gases to pass into the two branches known as the bronchi.

76. **(C)** The bronchi lead to the two lungs where they branch out in all directions into smaller tubules known as bronchioles.

77. **(D)** The vocal cords are found in the larynx.

78. **(B)** Air passes into the body via the nasal passage, then passes through the pharynx and on to the trachea.

79. **(A)** Insects (including bees) are within the phylum Arthropoda. Aves is the class composed of birds. Annelida is the phyla composed of segmented worms. Nematoda is the phyla of round worms. Porifera is the phyla of sponges.

80. **(B)** An enzyme is a protein, which is a polymer of amino acids. Enzymes generally have names ending in -*ase*—thus lactase is an enzyme. Lactose is the sugar that lactase acts upon. Glycogen, sucrose, and cellulose are all saccharides, not proteins.

81. **(A)** Vertebrates are divided into two main groups, the Aganatha (animals with no jaws) and the Gnathostomata (animals with jaws). Protista is a kingdom that includes algae and protozoa. Cnidaria is a phylum that contains jellyfish and hydra. Porifera is a phylum that contains sponges.

82. **(A)** The process whereby cells build molecules and store energy (in the form of covalent chemical bonds) is called anabolism. Catabolism is when molecules are broken apart. Adaptive radiation refers to an evolutionary

process. Lysis is a suffix meaning "to break apart." Inhibition occurs when a substance blocks the active site of an enzyme preventing the enzymatic reaction.

83. **(A)** Photolysis does not occur after glycolysis; rather, it is the first step in photosynthesis. Aerobic respiration occurs after glycolysis if oxygen is present in the cell. There are two steps in aerobic respiration for most organisms: the Krebs cycle (also known as the citric acid cycle) and electron transport. The first step, the Krebs cycle, occurs in the matrix of a cell's mitochondria and breaks down two pyruvic acid molecules into two CO_2 molecules, plus four H^+ (protons) and one molecule of ATP. The second step occurs along the electron transport system (or ETS) that captures the energy (in the form of electrons) that the Krebs cycle releases. If no oxygen is available, fermentation is the next step after glycolysis.

84. **(D)** The $\overset{\displaystyle O}{\underset{\displaystyle C\!-\!OH}{\|}}$ group is a carboxyl group and is the signature group found within organic acids.

85. **(B)** While microtubules, microfilaments, and centrioles all provide structure to cells of plants and animals, cell walls provide structure to plant cells (and some bacteria). Animal cells do not have a cell wall. Mitochondria are found in plant and animal cells, but function in cellular respiration, not in providing structure.

86. **(B)** Nonvascular plants are known as bryophytes (e.g., mosses). They lack tissue that will conduct water or food.

87. **(C)** Angiosperms are those plants that produce flowers as reproductive organs.

88. **(C)** There are two major divisions of angiosperms: monocots and dicots.

89. **(D)** Gymnosperms produce seeds without flowers.

90. **(D)** Gymnosperms include conifers (cone-bearers) and cycads.

91. **(A)** Lichens are often the first pioneer species to enter a rocky area after a volcanic eruption. Lichens break up the rock surfaces into soil, making the environment more hospitable for species of plants and animals. This represents an opportunistic life strategy or r-selection. Choices (B) and (D) represent equilibreal or k-selected life strategies that tend to remain long-term in a stable ecosystem. Choice (E) represents a short-term effect, not

a life-history strategy. Choice (C) may be a precursor to the invasion of an r-selected species.

92. **(B)** The cerebrum controls sensory and motor responses, and it controls memory, speech, and intelligence factors. It does not control involuntary muscles.

93. **(E)** While this statement is considered scientifically accurate, it is not a tenet of the cell theory. Choice (A) is a true statement about the cell theory, and choices (B–D) all represent tenets of the cell theory.

94. **(E)** Nitrogen is not absorbed into the ocean as part of the nitrogen cycle. Each of the other steps is included in the nitrogen cycle.

95. **(B)** Enzymes catalyze reactions in both living and nonliving environments.

96. **(B)** Vitamins are organic cofactors or coenzymes that are required by some enzymatic reactions. Vitamin C is required for collagen to be synthesized.

97. **(C)** The individual we recognize as an adult fern is actually the mature sporophyte. Both the mature gametophyte and the prothallus (which is the young gametophyte) are heart-shaped haploid structures that do not resemble an adult fern. The young sporophyte develops into the mature sporophyte, but while young it does not have the leaf structure characteristic of an adult fern.

98. **(E)** Tundra has extreme cold temperatures, low precipitation, modified grassland, permafrost, a short growing season, and some plants and animals.

99. **(D)** Desert has extreme hot or cold temperatures, with very low precipitation, sandy or rocky terrain, sparse vegetation (mainly succulents), small animals, rodents, and reptiles.

100. **(A)** Savanna is a kind of plain characterized by a warm climate, grassland, and seasonally dry climatic conditions.

101. **(B)** This response to a light stimulus is called phototropism. The hormone auxin, in response to the light, migrates from the light to the dark side of the shoot tip. The cells on the dark side now contain more auxin, which causes the cells on that side to elongate more rapidly than cells on the light side. The result is that the plant bends toward the light.

102. **(D)** The theory of punctuated equilibrium can account for the sudden appearance and disappearance of fossil species. The fossil record shows periods of stability with regard to appearance and disappearance of species as well as periods of sudden change.

103. **(B)** Hemophilia is a sex-linked recessive disease. Like color-blindness, the gene for hemophilia, h, is carried on the X-chromosome. If a male inherits the gene, he will have the genotype X^hY and will be a hemophiliac (a normal male is X^HY) since the recessive gene will be expressed. If a female inherits the gene, she will have the genotype X^HX^h and will carry the trait since her other X chromosome has the normal dominant gene, H.

   The common pattern of transmittal is from carrier mothers to their sons. Note that a carrier mother X^HX^h has an equal (50%) chance of passing the gene on to either a son (X^hY) or a daughter (X^HX^h); however, the daughter will not express the disease. It is unlikely for a female to be a hemophiliac, X^hX^h, since she must have acquired the recessive gene from both her carrier mother and her hemophiliac father. However, this is possible, and as expected, the incidence increases when there is marriage between relatives.

   If the gene were Y-linked, then a diseased father would always produce a hemophiliac son. However, a son inherits the gene only from his mother, since the mother contributes his sole X chromosome.

104. **(B)** The fungi encompass an entire kingdom in the classification scheme. They function as decomposers of organic matter and hence aid in the carbon, nitrogen, and phosphorus cycles. Of interest, fungi decompose both living and nonliving matter. The term *saprophytic* refers to its ability to decompose dead matter. This is in contrast to parasitic behavior, exhibited by some fungi, which refers to decomposition of living matter.

   All fungi are eukaryotic. The eukaryotes (literally, from the New Latin, "those having a true nucleus") include all organisms in kingdom Fungi, Animalia, Plantae, and Protista. They have a distinct nucleus enclosed in a membrane and many organelles. Only organisms of kingdom Monera (bacteria and cyanobacteria) are prokaryotic.

   Prokaryotes do not have a distinct nucleus. Rather their DNA is in a nucleoid region.

The term *heterotrophic* refers to the inability to manufacture one's own food. Fungi secrete digestive enzymes onto their food substrate and then absorb it. Animals are also heterotrophic, although animals ingest their food prior to digestion and absorption. In contrast, autotrophic organisms can produce their own food. For instance, plants produce food by photosynthesis. The monerans and protists are autotrophic or heterotrophic.

Some fungi are pathogenic, i.e., cause disease. Fungi can cause disease in animals (ringworm) and plants (potato blight). However, most fungi are not pathogenic and may even serve specific benefits for mankind. For instance, the antibiotic penicillin is produced by a fungus.

105. **(A)**   The pituitary gland is composed of an anterior and posterior lobe. The stalk of the posterior lobe is connected to the hypothalamus. Antidiuretic hormone (ADH) is produced in the hypothalamus and stored in the posterior pituitary. Upon nervous stimulation from the hypothalamus, the posterior pituitary releases ADH, which acts on kidney tubule to reabsorb water.

106. **(C)**   The pancreas secretes insulin to lower blood sugar and maintain equilibrium.

107. **(D)**   The adrenal glands produce adrenaline. This hormone is a well-known constrictor of blood vessels. The principle demonstrated here is that of negative feedback: a stimulus met by a response that reverses the trend of the stimulus. A dropping blood pressure must be opposed and corrected. Constricting blood vessels forces the same amount of blood to travel through a region of decreased volume; this causes a rise in blood pressure.

108. **(D)**   The hormone aldosterone is secreted by the adrenal cortex to promote sodium reabsorption in the kidney.

109. **(A)**   Restriction enzymes cleave strands of DNA segments at certain sites, thus yielding uniform fragments to be studied in the laboratory. The DNA molecule is not cleaved straight across by restriction enzymes; rather, these enzymes leave "sticky ends" that are complementary to another molecule cleaved by the same enzyme.

110. **(B)**   Of the choices given, carbon is the most abundant element found in protoplasm. Together with oxygen, hydrogen, and nitrogen, it composes over 90% of cellular structure. Calcium, phosphorus, and sulfur are found in varying amounts, depending upon the nature of the cell. Zinc is found in trace amounts.

111. **(C)** In photosynthesis, plant chloroplasts produce glucose and oxygen from carbon dioxide and water, in the presence of sunlight.

Glucose is a monosaccharide (simple sugar) that has many possible fates, depending upon the needs of the plant cell. Within the chloroplast, glucose can be polymerized into the polysaccharide starch. Alternatively, plant cells can respire: glucose can enter the cytoplasm and be oxidized/degraded to generate ATP. Cytoplasmic glucose can also participate in other chemical reactions, including the formation of the disaccharide sucrose (glucose + fructose) or other organic molecules.

Glycogen is the storage form of glucose in animal cells (liver and muscle cells). It is the analogue of plant starch, but is not found in plant cells.

112. **(C)** Phenylketonuria is an inherited disease in which the enzyme phenylalanine hydroxylase, which converts the amino acid phenylalanine to the amino acid tyrosine, is missing. Phenylalanine therefore is diverted to other usually insignificant metabolic pathways. These alternate metabolites develop to high levels in body fluids. Mental retardation and early death by age 20 are characteristic of the disease.

Like other inborn errors of metabolism, this disease is caused by an inherited lack of an enzyme (or synthesis of a deficient form of the enzyme), which leads to the accumulation of abnormal, or excessive levels of normal metabolites (metabolic intermediates). Symptoms may appear shortly after birth. There is no obvious "defect" at birth, since the disease is only manifest upon ingestion of the amino acid. Furthermore, birth defects may be environmentally or medically induced, rather than genetically induced.

Disorders of endocrine glands often include overproduction or lowered production of a hormone. For instance, a tumor of a gland may cause excess secretion. Underproduction may be a result of atrophy of the gland.

Nondisjunction refers to the event whereby the chromosomes do not separate during meiosis: the resultant gamete therefore carries an extra chromosome, or lacks a chromosome.

113. **(B)** Cells are classified as eukaryotic or prokaryotic. The former are characterized by a membrane-bound nucleus, while the latter do not have an organized nucleus. All living things are grouped into one of five kingdoms.

Only kingdom Monera consists of prokaryotic organisms. Monera include the cyanobacteria (blue-green algae) and the bacteria. E. coli (Escherichia coli) is a bacterium.

The other four kingdoms include only eukaryotic organisms. Kingdom Animalia, the animals, include the human being, or Homo sapiens. An oak tree falls into kingdom Plantae. Amoeba is a member of the kingdom Protista.

The AIDS (Acquired Immune Deficiency Syndrome) virus is not classified here because viruses are not truly living organisms; they depend on a living host (plant, animal, bacterium) for their metabolic and reproductive mechanisms.

114. **(A)**   Symbiosis, which literally means "living together," refers to an association between organisms of distinct species. When both organisms benefit from the association, the symbiosis is called mutualism. An extreme example of this is a lichen, which, through evolutionary time, is now a single organism. A lichen is a combination of a fungus and an alga. The alga supplies the photosynthetically produced food for the fungus while the fungus provides water and minerals, and a mechanical support for the alga.

In a commensalistic symbiosis, one species benefits, while the other is neither benefitted nor harmed. Epiphytes grow in the branches of trees to maximize light exposure, with no ill effect on the tree.

In parasitism, one species benefits at the expense of the other species, designated the host. Parasites may live on or within the host's body. A hookworm (phylum Nematoda) bores through human skin and eventually comes to live in the intestine, causing diarrhea and anemia in the host.

Members of a predatory species kill and consume other organisms in order to survive.

Competition occurs between organisms that share a limited resource in the environment. It can be intraspecific as well as interspecific.

115. **(E)** Charles Darwin is credited with formulating the most widely supported theory of evolution. The postulates of his theory came together in his book *On the Origin of Species by Means of Natural Selection* in 1859, and are recapitulated below.

All organisms overproduce gametes. Not all gametes form offspring, and of the offspring formed, not all survive. Those organisms that are most competitive (in various different aspects) will have greater likelihoods of survival. These survival traits vary from individual to individual but are passed on to the next generation, and thus over time, the best adaptions for survival are maintained. The environment determines which traits will be selected for or against; and these traits will change in time. A selected trait may later be disadvantageous.

The key drawback to Darwin's theory is that he did not suggest the key to variation in traits. It is now known that variation may be due to genetic mutations, gene flow due to migration, genetic drift (especially in small populations) and natural selection of genotypes, i.e., a differential ability to survive and/or reproduce.

APPENDIX

Laboratory Information

LABORATORY INFORMATION

MATHEMATICS, MEASUREMENT, AND DATA MANIPULATION

All branches of science rely heavily on the use of mathematics . . . whether it is the use of statistics in understanding populations and their impact on ecological factors, ratios and proportions when dealing with genetics, or calculus when considering forces and elasticity in physics. Mathematics is used for measuring and collecting data, interpreting and understanding the physical world, and comparing and presenting information about the physical world.

MEASUREMENTS AND ACCURACY

1. It is important to understand the difference between **precision** and **accuracy**. Accuracy represents how close a measurement is to the real or accepted value. Measurements are precise when multiple measurements are close to each other. For example, you might make an inaccurate standard solution for an acid-base titration. If the standard solution has a lower concentration than thought, more of it would be required to neutralize the sample, and you will think that the sample has a higher concentration. If you attempt the neutralization several times and get the same answer each time, the results would be precise, but not accurate.

2. Measurements in science are always taken in units of the International System of Units (the Système Internationale) or **SI Units**. Standard measurements in SI for volume, length, mass, etc., are in metric units.

Table A-1   Common SI Units

Unit	Unit Symbol	Quantity To Measure
meter	m	length
kilogram	kg	mass
second	s	time
degrees Celsius	°C	standard temperature
liter	l	volume
ampere	A	electric current
Kelvin	K	thermodynamic temperature
candela	cd	luminous intensity
mole	mol	amount of substance

Fig. A-1   Celsius (left) / fahrenheit (right) thermometer showing freezing point of water at 1 ATM pressure (0°C = 32°F).

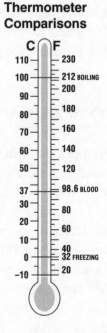

3. Be aware of **significant figures** in laboratory measurements. You are allowed one uncertain figure in your measurement. For example, in the following graduated cylinder measurement, the volume is delineated by tenths of a milliliter (mL).

a. First, remember to read the bottom of the **meniscus** when mea-
    suring volume in glassware. The meniscus is the bottom (or top)
    of the curve of the liquid in a measuring container. In Figure A-2
    below, the liquid would measure 17.0 mL, not 18.0 mL.

Fig. A-2 Graduated cylinder showing meniscus.

b. Since one uncertain digit is allowed, in the example (Figure A-3)
    below you should record 11.45 mL. It might be 11.44 mL; you are
    uncertain about the +/– 0.01 mL amount, but you are certain that the
    +/– 0.1 mL aspect of the measurement is between 11.4 and 11.5 mL.

Fig. A-3 Graduated cylinder measurement.

c. Also with regard to significant figures, remember that if they
    are multiplied or divided together to get some final answer, the
    number of significant figures in the answer should match the
    least number of significant figures in the numbers that lead to the
    answer.

Fig. A-4 Some standard laboratory equipment.

Erlenmeyer Flask

Beaker

Funnel

Buchner Funnel

Bunsen Burner

Crucible

Clamp and Ring
Stand

Distillation Apparatus

Burette

Filtering Flask

Hot Plate

Mortar and Pestle

Separatory Funnel

Thermometer

Volumetric Flask

Interpreting Results

1. Never expect quantitative measurements to exactly match the expected value. Be sure to calculate the **percent accuracy** (or conversely **percent error**) by dividing the difference between the measured and accepted/expected values, by the accepted/expected value, and multiplying by 100. This figure should be a part of any laboratory report involving quantitative measurements.

$$\% \text{ Error} = \left(\frac{|\text{ Your Measured Result} - \text{Accepted Value}|}{\text{Accepted Value}} \right) \times 100$$

2. Analyze the precision of your measurements by making multiple measurements or comparing your value with your classmates.

3. Use tables or graphs to better summarize multiple data points.

4. As you examine the accuracy of your measurements, consider each step that you undertook to make the measurement. Consider how an inadvertent error in that step could have influenced the measurement, either by increasing or decreasing the value of the measurement relative to the expected value.

Reporting Results

1. When reporting data, use an appropriate form of notation for the discipline of the experiment. For example, certain chemistry experiments may require notation in chemical equations and molecular diagrams. Other experimental results, particularly those that are very large or small in scale, should be reported in scientific notation. Scientific notation is accomplished by writing any number in the form . . . $a \times 10^b$ . . . where the number is expressed as a power of ten such that a is equivalent to a number between 1 and 10.

Examples: $2{,}398 = 2.398 \times 10^3$ in scientific notation
$0.00000626 = 6.26 \times 10^{-6}$
$-42 = -4.2 \times 10^1$

2. Some data reports will need to be in the form of graphs, ratios, percentages, or some other form. Always be sure to check what form is required for any type of data display or report and be sure to include appropriate units of measure where applicable (gallons, Newtons, pounds, PSI, etc.).

When graphing a value as a function of another value (variable), the value that you measure should be on the *y*-axis; this is the dependent variable. It has changed as a result of changing the independent variable, which should be on the *x*-axis. In other words, the *y*-axis (the dependent variable) depends on *x* (the independent variable) which "just is"—it doesn't depend on anything within the experiment. In the following example, we have used *Days 1–10* as the independent variable (*x*) and *Temperature* measured on those days as the dependent variable (*y*). The temperature "depended" on which day it was. We put the measurements first in a table, then graphed them.

Table A-2   Day versus Temperature

First Coordinate, the Independent Variable, In this example: Day	Second Coordinate, the Dependent Variable In this example: Temperature in Degrees
1	–7.6
2	–2.0
3	5.2
4	–1.0
5	–5.7
6	6.0
7	6.6
8	18.0
9	5.6
10	5.8

Fig. A-5   Graph of Day versus Temperature.

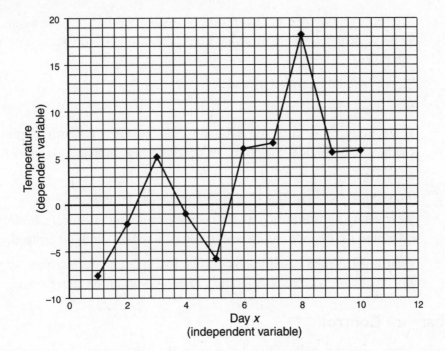

LABORATORY PROCEDURES AND SAFETY

One of the essential activities of biology is laboratory work. It is important to understand the safe and appropriate use of all laboratory apparatus, models, and specimens.

The following safety guidelines are minimum expectations for a safe laboratory experience:

General

- Keep laboratory and storage areas neat and clean. Accidents are more likely to occur in a disheveled lab or with items improperly labeled or cared for.

- Be sure all laboratory participants are aware of safety procedures for all activities, equipment, and chemicals.

Apparel

- Clothing should not be baggy or have loosely hanging items such as belts, scarves, or jewelry that could catch on fire or get caught in equipment.

- Sleeves should be rolled up and secured.

- Hair should be tied back.

- Wear goggles during all procedures involving moving objects, chemicals, biological materials, or anything else indicating a need for goggles.

- Shoes must be close-toed to protect from spills or other hazards.

- Wear safety aprons during all procedures involving chemicals, biological materials, or anything else indicating a need for clothing protection.

- Wear gloves when handling chemicals, live specimens, microbes, or in any other case where the laboratory instructions call for glove use.

Substance Control

- When working with any substance in the laboratory, be particularly aware of product labels and instructions—follow all safety and disposal guidelines. Keep all products in original containers or in an appropriately labeled container that is suitable for that product.

- Never eat or drink anything in a laboratory situation.

- Be aware of Material Safety Data Sheets (MSDSs) for any and all substances used in a laboratory activity.

- Be familiar with safety equipment location and use (such as eye-wash station, emergency cutoff valves, spill kits, etc.).

- Pay special attention to harmful and corrosive liquids and gases. Use venting appropriately. Always transfer acids or bases into water, stirring (not water into an acid or a base).

- Keep all poisonous, corrosive, flammable, and otherwise dangerous substances (check MSDS sheets) in a locked cabinet when not in use.

- Dispose of substances and specimens according to supplier instructions and all state, local, and laboratory policies.

Fire Control

- Know the location and proper use protocol of a fire blanket, fire extinguishers, emergency shutoff buttons, etc.
- Instruct students in safe fire exit practices.

Accidents

- Report all accidents according to laboratory policy.
- Clean up all spills and broken objects according to the supplier's directions.

Handling Live Organisms

- No investigations in the laboratory should inflict pain or harm on a mammal, bird, reptile, fish, or amphibian.
- Treat all living things, including plants and invertebrates with appropriate care and respect.
- Use antiseptic procedures with all microbes as if all are dangerous.

ANSWER SHEETS

Practice Test 1
Practice Test 2

PRACTICE TEST 1

Answer Sheet

1. Ⓐ Ⓑ Ⓒ Ⓓ Ⓔ	40. Ⓐ Ⓑ Ⓒ Ⓓ Ⓔ	79. Ⓐ Ⓑ Ⓒ Ⓓ Ⓔ
2. Ⓐ Ⓑ Ⓒ Ⓓ Ⓔ	41. Ⓐ Ⓑ Ⓒ Ⓓ Ⓔ	80. Ⓐ Ⓑ Ⓒ Ⓓ Ⓔ
3. Ⓐ Ⓑ Ⓒ Ⓓ Ⓔ	42. Ⓐ Ⓑ Ⓒ Ⓓ Ⓔ	81. Ⓐ Ⓑ Ⓒ Ⓓ Ⓔ
4. Ⓐ Ⓑ Ⓒ Ⓓ Ⓔ	43. Ⓐ Ⓑ Ⓒ Ⓓ Ⓔ	82. Ⓐ Ⓑ Ⓒ Ⓓ Ⓔ
5. Ⓐ Ⓑ Ⓒ Ⓓ Ⓔ	44. Ⓐ Ⓑ Ⓒ Ⓓ Ⓔ	83. Ⓐ Ⓑ Ⓒ Ⓓ Ⓔ
6. Ⓐ Ⓑ Ⓒ Ⓓ Ⓔ	45. Ⓐ Ⓑ Ⓒ Ⓓ Ⓔ	84. Ⓐ Ⓑ Ⓒ Ⓓ Ⓔ
7. Ⓐ Ⓑ Ⓒ Ⓓ Ⓔ	46. Ⓐ Ⓑ Ⓒ Ⓓ Ⓔ	85. Ⓐ Ⓑ Ⓒ Ⓓ Ⓔ
8. Ⓐ Ⓑ Ⓒ Ⓓ Ⓔ	47. Ⓐ Ⓑ Ⓒ Ⓓ Ⓔ	86. Ⓐ Ⓑ Ⓒ Ⓓ Ⓔ
9. Ⓐ Ⓑ Ⓒ Ⓓ Ⓔ	48. Ⓐ Ⓑ Ⓒ Ⓓ Ⓔ	87. Ⓐ Ⓑ Ⓒ Ⓓ Ⓔ
10. Ⓐ Ⓑ Ⓒ Ⓓ Ⓔ	49. Ⓐ Ⓑ Ⓒ Ⓓ Ⓔ	88. Ⓐ Ⓑ Ⓒ Ⓓ Ⓔ
11. Ⓐ Ⓑ Ⓒ Ⓓ Ⓔ	50. Ⓐ Ⓑ Ⓒ Ⓓ Ⓔ	89. Ⓐ Ⓑ Ⓒ Ⓓ Ⓔ
12. Ⓐ Ⓑ Ⓒ Ⓓ Ⓔ	51. Ⓐ Ⓑ Ⓒ Ⓓ Ⓔ	90. Ⓐ Ⓑ Ⓒ Ⓓ Ⓔ
13. Ⓐ Ⓑ Ⓒ Ⓓ Ⓔ	52. Ⓐ Ⓑ Ⓒ Ⓓ Ⓔ	91. Ⓐ Ⓑ Ⓒ Ⓓ Ⓔ
14. Ⓐ Ⓑ Ⓒ Ⓓ Ⓔ	53. Ⓐ Ⓑ Ⓒ Ⓓ Ⓔ	92. Ⓐ Ⓑ Ⓒ Ⓓ Ⓔ
15. Ⓐ Ⓑ Ⓒ Ⓓ Ⓔ	54. Ⓐ Ⓑ Ⓒ Ⓓ Ⓔ	93. Ⓐ Ⓑ Ⓒ Ⓓ Ⓔ
16. Ⓐ Ⓑ Ⓒ Ⓓ Ⓔ	55. Ⓐ Ⓑ Ⓒ Ⓓ Ⓔ	94. Ⓐ Ⓑ Ⓒ Ⓓ Ⓔ
17. Ⓐ Ⓑ Ⓒ Ⓓ Ⓔ	56. Ⓐ Ⓑ Ⓒ Ⓓ Ⓔ	95. Ⓐ Ⓑ Ⓒ Ⓓ Ⓔ
18. Ⓐ Ⓑ Ⓒ Ⓓ Ⓔ	57. Ⓐ Ⓑ Ⓒ Ⓓ Ⓔ	96. Ⓐ Ⓑ Ⓒ Ⓓ Ⓔ
19. Ⓐ Ⓑ Ⓒ Ⓓ Ⓔ	58. Ⓐ Ⓑ Ⓒ Ⓓ Ⓔ	97. Ⓐ Ⓑ Ⓒ Ⓓ Ⓔ
20. Ⓐ Ⓑ Ⓒ Ⓓ Ⓔ	59. Ⓐ Ⓑ Ⓒ Ⓓ Ⓔ	98. Ⓐ Ⓑ Ⓒ Ⓓ Ⓔ
21. Ⓐ Ⓑ Ⓒ Ⓓ Ⓔ	60. Ⓐ Ⓑ Ⓒ Ⓓ Ⓔ	99. Ⓐ Ⓑ Ⓒ Ⓓ Ⓔ
22. Ⓐ Ⓑ Ⓒ Ⓓ Ⓔ	61. Ⓐ Ⓑ Ⓒ Ⓓ Ⓔ	100. Ⓐ Ⓑ Ⓒ Ⓓ Ⓔ
23. Ⓐ Ⓑ Ⓒ Ⓓ Ⓔ	62. Ⓐ Ⓑ Ⓒ Ⓓ Ⓔ	101. Ⓐ Ⓑ Ⓒ Ⓓ Ⓔ
24. Ⓐ Ⓑ Ⓒ Ⓓ Ⓔ	63. Ⓐ Ⓑ Ⓒ Ⓓ Ⓔ	102. Ⓐ Ⓑ Ⓒ Ⓓ Ⓔ
25. Ⓐ Ⓑ Ⓒ Ⓓ Ⓔ	64. Ⓐ Ⓑ Ⓒ Ⓓ Ⓔ	103. Ⓐ Ⓑ Ⓒ Ⓓ Ⓔ
26. Ⓐ Ⓑ Ⓒ Ⓓ Ⓔ	65. Ⓐ Ⓑ Ⓒ Ⓓ Ⓔ	104. Ⓐ Ⓑ Ⓒ Ⓓ Ⓔ
27. Ⓐ Ⓑ Ⓒ Ⓓ Ⓔ	66. Ⓐ Ⓑ Ⓒ Ⓓ Ⓔ	105. Ⓐ Ⓑ Ⓒ Ⓓ Ⓔ
28. Ⓐ Ⓑ Ⓒ Ⓓ Ⓔ	67. Ⓐ Ⓑ Ⓒ Ⓓ Ⓔ	106. Ⓐ Ⓑ Ⓒ Ⓓ Ⓔ
29. Ⓐ Ⓑ Ⓒ Ⓓ Ⓔ	68. Ⓐ Ⓑ Ⓒ Ⓓ Ⓔ	107. Ⓐ Ⓑ Ⓒ Ⓓ Ⓔ
30. Ⓐ Ⓑ Ⓒ Ⓓ Ⓔ	69. Ⓐ Ⓑ Ⓒ Ⓓ Ⓔ	108. Ⓐ Ⓑ Ⓒ Ⓓ Ⓔ
31. Ⓐ Ⓑ Ⓒ Ⓓ Ⓔ	70. Ⓐ Ⓑ Ⓒ Ⓓ Ⓔ	109. Ⓐ Ⓑ Ⓒ Ⓓ Ⓔ
32. Ⓐ Ⓑ Ⓒ Ⓓ Ⓔ	71. Ⓐ Ⓑ Ⓒ Ⓓ Ⓔ	110. Ⓐ Ⓑ Ⓒ Ⓓ Ⓔ
33. Ⓐ Ⓑ Ⓒ Ⓓ Ⓔ	72. Ⓐ Ⓑ Ⓒ Ⓓ Ⓔ	111. Ⓐ Ⓑ Ⓒ Ⓓ Ⓔ
34. Ⓐ Ⓑ Ⓒ Ⓓ Ⓔ	73. Ⓐ Ⓑ Ⓒ Ⓓ Ⓔ	112. Ⓐ Ⓑ Ⓒ Ⓓ Ⓔ
35. Ⓐ Ⓑ Ⓒ Ⓓ Ⓔ	74. Ⓐ Ⓑ Ⓒ Ⓓ Ⓔ	113. Ⓐ Ⓑ Ⓒ Ⓓ Ⓔ
36. Ⓐ Ⓑ Ⓒ Ⓓ Ⓔ	75. Ⓐ Ⓑ Ⓒ Ⓓ Ⓔ	114. Ⓐ Ⓑ Ⓒ Ⓓ Ⓔ
37. Ⓐ Ⓑ Ⓒ Ⓓ Ⓔ	76. Ⓐ Ⓑ Ⓒ Ⓓ Ⓔ	115. Ⓐ Ⓑ Ⓒ Ⓓ Ⓔ
38. Ⓐ Ⓑ Ⓒ Ⓓ Ⓔ	77. Ⓐ Ⓑ Ⓒ Ⓓ Ⓔ	
39. Ⓐ Ⓑ Ⓒ Ⓓ Ⓔ	78. Ⓐ Ⓑ Ⓒ Ⓓ Ⓔ	

PRACTICE TEST 2

Answer Sheet

1. Ⓐ Ⓑ Ⓒ Ⓓ Ⓔ	40. Ⓐ Ⓑ Ⓒ Ⓓ Ⓔ	79. Ⓐ Ⓑ Ⓒ Ⓓ Ⓔ
2. Ⓐ Ⓑ Ⓒ Ⓓ Ⓔ	41. Ⓐ Ⓑ Ⓒ Ⓓ Ⓔ	80. Ⓐ Ⓑ Ⓒ Ⓓ Ⓔ
3. Ⓐ Ⓑ Ⓒ Ⓓ Ⓔ	42. Ⓐ Ⓑ Ⓒ Ⓓ Ⓔ	81. Ⓐ Ⓑ Ⓒ Ⓓ Ⓔ
4. Ⓐ Ⓑ Ⓒ Ⓓ Ⓔ	43. Ⓐ Ⓑ Ⓒ Ⓓ Ⓔ	82. Ⓐ Ⓑ Ⓒ Ⓓ Ⓔ
5. Ⓐ Ⓑ Ⓒ Ⓓ Ⓔ	44. Ⓐ Ⓑ Ⓒ Ⓓ Ⓔ	83. Ⓐ Ⓑ Ⓒ Ⓓ Ⓔ
6. Ⓐ Ⓑ Ⓒ Ⓓ Ⓔ	45. Ⓐ Ⓑ Ⓒ Ⓓ Ⓔ	84. Ⓐ Ⓑ Ⓒ Ⓓ Ⓔ
7. Ⓐ Ⓑ Ⓒ Ⓓ Ⓔ	46. Ⓐ Ⓑ Ⓒ Ⓓ Ⓔ	85. Ⓐ Ⓑ Ⓒ Ⓓ Ⓔ
8. Ⓐ Ⓑ Ⓒ Ⓓ Ⓔ	47. Ⓐ Ⓑ Ⓒ Ⓓ Ⓔ	86. Ⓐ Ⓑ Ⓒ Ⓓ Ⓔ
9. Ⓐ Ⓑ Ⓒ Ⓓ Ⓔ	48. Ⓐ Ⓑ Ⓒ Ⓓ Ⓔ	87. Ⓐ Ⓑ Ⓒ Ⓓ Ⓔ
10. Ⓐ Ⓑ Ⓒ Ⓓ Ⓔ	49. Ⓐ Ⓑ Ⓒ Ⓓ Ⓔ	88. Ⓐ Ⓑ Ⓒ Ⓓ Ⓔ
11. Ⓐ Ⓑ Ⓒ Ⓓ Ⓔ	50. Ⓐ Ⓑ Ⓒ Ⓓ Ⓔ	89. Ⓐ Ⓑ Ⓒ Ⓓ Ⓔ
12. Ⓐ Ⓑ Ⓒ Ⓓ Ⓔ	51. Ⓐ Ⓑ Ⓒ Ⓓ Ⓔ	90. Ⓐ Ⓑ Ⓒ Ⓓ Ⓔ
13. Ⓐ Ⓑ Ⓒ Ⓓ Ⓔ	52. Ⓐ Ⓑ Ⓒ Ⓓ Ⓔ	91. Ⓐ Ⓑ Ⓒ Ⓓ Ⓔ
14. Ⓐ Ⓑ Ⓒ Ⓓ Ⓔ	53. Ⓐ Ⓑ Ⓒ Ⓓ Ⓔ	92. Ⓐ Ⓑ Ⓒ Ⓓ Ⓔ
15. Ⓐ Ⓑ Ⓒ Ⓓ Ⓔ	54. Ⓐ Ⓑ Ⓒ Ⓓ Ⓔ	93. Ⓐ Ⓑ Ⓒ Ⓓ Ⓔ
16. Ⓐ Ⓑ Ⓒ Ⓓ Ⓔ	55. Ⓐ Ⓑ Ⓒ Ⓓ Ⓔ	94. Ⓐ Ⓑ Ⓒ Ⓓ Ⓔ
17. Ⓐ Ⓑ Ⓒ Ⓓ Ⓔ	56. Ⓐ Ⓑ Ⓒ Ⓓ Ⓔ	95. Ⓐ Ⓑ Ⓒ Ⓓ Ⓔ
18. Ⓐ Ⓑ Ⓒ Ⓓ Ⓔ	57. Ⓐ Ⓑ Ⓒ Ⓓ Ⓔ	96. Ⓐ Ⓑ Ⓒ Ⓓ Ⓔ
19. Ⓐ Ⓑ Ⓒ Ⓓ Ⓔ	58. Ⓐ Ⓑ Ⓒ Ⓓ Ⓔ	97. Ⓐ Ⓑ Ⓒ Ⓓ Ⓔ
20. Ⓐ Ⓑ Ⓒ Ⓓ Ⓔ	59. Ⓐ Ⓑ Ⓒ Ⓓ Ⓔ	98. Ⓐ Ⓑ Ⓒ Ⓓ Ⓔ
21. Ⓐ Ⓑ Ⓒ Ⓓ Ⓔ	60. Ⓐ Ⓑ Ⓒ Ⓓ Ⓔ	99. Ⓐ Ⓑ Ⓒ Ⓓ Ⓔ
22. Ⓐ Ⓑ Ⓒ Ⓓ Ⓔ	61. Ⓐ Ⓑ Ⓒ Ⓓ Ⓔ	100. Ⓐ Ⓑ Ⓒ Ⓓ Ⓔ
23. Ⓐ Ⓑ Ⓒ Ⓓ Ⓔ	62. Ⓐ Ⓑ Ⓒ Ⓓ Ⓔ	101. Ⓐ Ⓑ Ⓒ Ⓓ Ⓔ
24. Ⓐ Ⓑ Ⓒ Ⓓ Ⓔ	63. Ⓐ Ⓑ Ⓒ Ⓓ Ⓔ	102. Ⓐ Ⓑ Ⓒ Ⓓ Ⓔ
25. Ⓐ Ⓑ Ⓒ Ⓓ Ⓔ	64. Ⓐ Ⓑ Ⓒ Ⓓ Ⓔ	103. Ⓐ Ⓑ Ⓒ Ⓓ Ⓔ
26. Ⓐ Ⓑ Ⓒ Ⓓ Ⓔ	65. Ⓐ Ⓑ Ⓒ Ⓓ Ⓔ	104. Ⓐ Ⓑ Ⓒ Ⓓ Ⓔ
27. Ⓐ Ⓑ Ⓒ Ⓓ Ⓔ	66. Ⓐ Ⓑ Ⓒ Ⓓ Ⓔ	105. Ⓐ Ⓑ Ⓒ Ⓓ Ⓔ
28. Ⓐ Ⓑ Ⓒ Ⓓ Ⓔ	67. Ⓐ Ⓑ Ⓒ Ⓓ Ⓔ	106. Ⓐ Ⓑ Ⓒ Ⓓ Ⓔ
29. Ⓐ Ⓑ Ⓒ Ⓓ Ⓔ	68. Ⓐ Ⓑ Ⓒ Ⓓ Ⓔ	107. Ⓐ Ⓑ Ⓒ Ⓓ Ⓔ
30. Ⓐ Ⓑ Ⓒ Ⓓ Ⓔ	69. Ⓐ Ⓑ Ⓒ Ⓓ Ⓔ	108. Ⓐ Ⓑ Ⓒ Ⓓ Ⓔ
31. Ⓐ Ⓑ Ⓒ Ⓓ Ⓔ	70. Ⓐ Ⓑ Ⓒ Ⓓ Ⓔ	109. Ⓐ Ⓑ Ⓒ Ⓓ Ⓔ
32. Ⓐ Ⓑ Ⓒ Ⓓ Ⓔ	71. Ⓐ Ⓑ Ⓒ Ⓓ Ⓔ	110. Ⓐ Ⓑ Ⓒ Ⓓ Ⓔ
33. Ⓐ Ⓑ Ⓒ Ⓓ Ⓔ	72. Ⓐ Ⓑ Ⓒ Ⓓ Ⓔ	111. Ⓐ Ⓑ Ⓒ Ⓓ Ⓔ
34. Ⓐ Ⓑ Ⓒ Ⓓ Ⓔ	73. Ⓐ Ⓑ Ⓒ Ⓓ Ⓔ	112. Ⓐ Ⓑ Ⓒ Ⓓ Ⓔ
35. Ⓐ Ⓑ Ⓒ Ⓓ Ⓔ	74. Ⓐ Ⓑ Ⓒ Ⓓ Ⓔ	113. Ⓐ Ⓑ Ⓒ Ⓓ Ⓔ
36. Ⓐ Ⓑ Ⓒ Ⓓ Ⓔ	75. Ⓐ Ⓑ Ⓒ Ⓓ Ⓔ	114. Ⓐ Ⓑ Ⓒ Ⓓ Ⓔ
37. Ⓐ Ⓑ Ⓒ Ⓓ Ⓔ	76. Ⓐ Ⓑ Ⓒ Ⓓ Ⓔ	115. Ⓐ Ⓑ Ⓒ Ⓓ Ⓔ
38. Ⓐ Ⓑ Ⓒ Ⓓ Ⓔ	77. Ⓐ Ⓑ Ⓒ Ⓓ Ⓔ	
39. Ⓐ Ⓑ Ⓒ Ⓓ Ⓔ	78. Ⓐ Ⓑ Ⓒ Ⓓ Ⓔ	

Glossary

abiotic: nonliving factors that support an organism's life and reproduction within its habitat

active transport: transport across cell membrane with energy output from the cell

alleles: different forms of corresponding genes on matching chromosomes

altruism: social behavior of organisms to tend to serve the needs of the society as a whole in addition to its own individual needs

amensalism: type of symbiosis where one species is neither helped nor harmed while one species' growth is inhibited

anabolism: process of cells building molecules and storing energy (in the form of chemical bonds)

antibodies: protein produced by the immune system; find and attach to foreign toxins (antigen) to mark it for destruction

antigen: toxins, bacteria, foreign cells, etc. that enter the body and are fought by the immune system

amylase: starch-digesting enzyme

amino acids: compounds of carbon, hydrogen, oxygen, nitrogen, and sometimes sulfur, phosphorous that combine in various sequences to form proteins

ATP: adenosine triphosphate; energy currency of cellular activity; efficient storage molecule for energy needed for cellular processes; nitrogenous base (adenine), a simple sugar (ribose), and three phosphate group

autotrophs: first trophic level, producers; produce their own food through photosynthesis

bile: chemical produced by liver that aids in digesting fats and carries away broken-down pigments and chemicals

biomass: total mass of a species in an ecosystem

biotic: living factors that support an organs life and reproduction within its habitat

capillaries: smallest vessels that surround all tissues of the body and exchange carbon dioxide for oxygen

catabolism: process of breaking down molecules and releasing stored energy

catalyst: a substance that changes the speed of a reaction without being affected itself; enzyme names have the suffix *ase* (i.e., polymerase, lactase)

cell membrane: membrane encasing cell; composed of a double layer (bilayer) of phospholipids with globular proteins embedded within the layers

cell wall: cellulose and lignin enclosure of plant and some bacterial cells

central vacuole: fluid-filled space that stores water and soluble nutrients for the plant's use

centrioles: tubes constructed of a geometrical arrangement of microtubules in a pinwheel shape; function to form new microtubules and structural skeleton around which cells split during mitosis and meiosis

cerebellum: portion of brain that controls balance, equilibrium, and muscle coordination

cerebrum: section of brain that controls sensory and motor responses, memory, speech, and most factors of intelligence

chlorophyll: pigment molecules that give the chloroplast their green color and where photosynthesis occurs

chloroplasts: organelles of photosynthesis in plant cells (and some protists)

247

chromosomes: independent strands of DNA that carry all the genetic information for a given organism

conditioning: learned behavior to apply an old response to a new stimulus; example, Pavlov's dogs

cuticle: waxy leaf covering; maintains leaf's moisture balance

cytoskeleton: organelle that provides structural support to a cell

diffusion: process of ion flow through the cell membrane from area of higher concentration to area of lower concentration (tending to equalize concentrations)

digestion: breaking down ingested particles into molecules that can be absorbed by the body

dihybrid cross: cross between two individuals considering two separate traits

dominant allele: masks the effect of its partner allele

endocytic vesicles: sack of plasma membrane that surrounds a molecule outside the membrane and brings a molecule or substance into the cytoplasm

endocytosis: process where large molecules (e.g., sugars, proteins, etc.) are taken up into a pocket of membrane, pinched off, and delivered inside a membrane sack into the cytoplasm

endoplasmic reticulum (ER): organelle consisting of folded membranes; location of processing and delivery of lipids, proteins, and steroids

enzymes: protein molecules that act as catalysts for organic reactions

eukaryotes: cells with membrane-bound intracellular organelles, including a nucleus and DNA organized into chromosomes

exocytosis: process where large molecules (e.g., sugars, proteins, etc.) are taken up into a pocket of membrane, pinched off, and delivered outside the cell membrane enzyme; protein that acts as a catalyst for reactions

egg cell: female gamete

egestion: elimination of indigestible materials from the body at the end of the digestion process

embryo: baby organism; growing zygote

epidermis: outermost layer of plant leaf; generally one cell thick; secretes waxy cuticle and protects the inner tissue of the leaf

facilitated diffusion: movement across cell membrane with assistance of specialized proteins

fats: highly efficient lipid molecules used for long-term energy storage; energy is stored in chemical bonds between the atoms of lipid molecules, when bonds are broken, energy is released; in addition to storing energy, fats also function in organisms to provide a protective layer that insulates internal organs and maintains heat within the body

fixed action patterns (FAP): complex but stereotyped behaviors in response to a stimulus.

gametes: resultant haploid cells of meiosis (egg and sperm)

gastrointestinal (GI) tract: organs of digestion; includes mouth, pharynx, esophagus, stomach, small intestine, large intestine, rectum, and anus in mammals plus accessory organs teeth, tongue, salivary glands, liver, gallbladder, and pancreas

gene: length of DNA that encodes a particular protein

genome: sum total of genetic information of many organisms, including humans

genotype: combination of alleles that make a particular trait

Golgi apparatus (body, complex): organelle that stores, packages, and ships proteins

glycogen: polysaccharide composed of many joined glucose units; many animals use glycogen as a short-term storage molecule for energy, in mammals, glycogen is found in muscle and liver tissue

habitat: physical place where a species lives

habituation: learned behavior where the organism produces less and less response as a stimulus is repeated, without a subsequent negative or positive action

hemoglobin: protein that carries iron in red blood cells

homeostasis: condition of an organism; all systems are within acceptable healthy range

homolog: matching pair of homologous chromosomes

homologous chromosomes: set of matching chromosomes with analogous DNA

hormones: chemicals produced in endocrine glands that travel through the circulatory system, are taken up by specific targeted organs or tissues and modify metabolic activities

hypothalamus: organ within brain involved in hunger, thirst, blood pressure, body temperature, hostility, pain, pleasure, etc.

ingestion: intake of food to the gastrointestinal tract

instincts: innate highly stereotyped behaviors

invertebrate: animal species having no internal backbone structure

kineses: changes in speed of movement in response to stimuli

lipids: are organic compounds composed of carbon, hydrogen, and oxygen with ratio of hydrogen to oxygen always greater than 2:1; include waxes, steroids, phospholipids, and fats

lymph: collection of excess fluid and plasma proteins absorbed from between cells into a special system of vessels

lymph nodes: small masses of lymph tissue; filter lymph and produce lymphocytes

lysosome: membrane-bound organelles containing digestive enzymes; digest unused material within the cell, damaged organelles, or materials absorbed by the cell for use

medulla oblongata: portion of brain that controls involuntary responses such as breathing and heartbeat

meiosis: process of producing four haploid (single chromosome gamete cells) from one parent diploid (double chromosome) cell

microfilaments: double-stranded chains of proteins, give structure to the cell

microtubules: long, hollow, cylindrical protein filaments that give structural support to a cell; also found at base of flagella

microvilli: projections of a cell extending from the cell membrane; filaments that increase surface area of cell membrane in certain types of cells, increasing the area available to absorb nutrients; also contain enzymes involved in digesting certain types of nutrients

mitochondria: organelles of cellular respiration (the process of breaking up covalent bonds within sugar molecules with the intake of oxygen and release of ATP, adenosine tri-phosphate)

mitosis: asexual reproduction of cells from one parent cell to two nearly identical daughter cells each with a full set of chromosomes

monohybrid cross: a cross between two individuals where only one trait is considered.

mutation: DNA copying error that randomly occurs during replication of or resulting from damage to DNA caused by environmental factors

mutualism: form of symbiosis where both species benefit

myelin sheath: insulation along axon of nerve cells that speeds electrochemical conduction

neurons: nerve cells; carry impulses via electrochemical responses through cell body and axon (long root-like appendage of the cell)

niche: the role a species plays within an ecosystem; includes physical requirements (such as light and water) and biological activities (how it reproduces, how it acquires food, etc.)

nuclear membrane: phospholipid bilayer boundary between the nucleus and the cytoplasm

nuclear pores: points at which the double nuclear membrane fuses together forming a passageway between the inside of the nucleus and the cytoplasm outside the nucleus allowing the cell to selectively move molecules in and out of the nucleus

nucleic acids: compounds composed of chains of nucleotides; including RNA (ribonucleic acid), DNA (deoxyribonucleic acid)

nucleotides: monomers that form nucleic acids; each nucleotide has a sugar (from the pentose group) attached to a phosphate group and a nitrogenous base

nucleolus: rounded area within the nucleus of the cell where ribosomal RNA is synthesized

nucleus: central cell organelle surrounded by two lipid bilayer membranes; contains chromosomes, nuclear pores, nucleoplasm, and nucleolus

osmosis: process of diffusion occurring only with water molecules; requires no addition of energy, occurs when water concentration inside the cell differs from concentration outside the cell

parasitism: symbiosis in which one species benefits, but the other is harmed

passive transport: movement of substances across cell membrane without the cell expending energy

phenotype: trait expressed by a particular combination of alleles (genotype)

phloem: stem tissue of plants composed of long tubular cells tissue; made of stacked cells connected by sieve plates (which allow nutrients to pass from cell to cell); transports food made in the leaves (by photosynthesis) to the rest of the plant

photosynthesis: set of reactions that convert the light energy of the sun into chemical energy usable by living things;

$$6CO_2 + 6H_2O + \text{light energy} \rightarrow C_6H_2O_6 + 6O_2$$

(carbon dioxide + water → glucose + oxygen)

pituitary gland: gland within the brain that releases various hormones for bodily functions

placenta: connection between the mother and embryo in mammals; site of transfer for nutrients, water, and wastes

predator: an organism that eats another

prey: organism eaten by another

prokaryotes: cells with no nucleus or any other membrane-bound organelles (cell components that perform particular functions)

proteins: large un-branched chains of amino acids

Punnett square: notation that easily predicts results of a genetic cross

recessive: allele that does not produce its trait when present with a dominant allele

reflexes: automatic movement of a body part in response to a stimulus

regulatory genes: genes that code proteins that determine functional or physiological events, such as growth

ribosomes: site of protein synthesis within cells; composed of protein molecules and RNA

rough endoplasmic reticulum (RER): endoplasmic reticulum with attached ribosomes

ruminants: mammals that consume large amounts of vegetation have several chambers in their stomachs and regurgitate food from the first two stomach chambers to rechew as cud; allowing much of their food to be broken down mechanically

secretory vesicles: packets of substances produced within the cell (e.g., protein) carried to cell membrane

sporophyte: diploid generation in spore producing plants

sperm: male gamete

stomata: openings on underside of leaves ringed by guard cells; allow moisture and gases (carbon dioxide and oxygen) to pass in and out of the leaf, facilitating photosynthesis

structural genes: genes that code proteins that form organs and structural characteristics

symbiosis: when two species interact with each other within the same range

taxes (plural of taxis): directional responses either toward or away from a stimulus

thalamus: organ within brain that integrates sensory information in the brain

transcription: formation of an RNA molecule corresponding to a gene

transpiration: evaporation of water from leaves of plants; siphoning effect that continues to pull water up from the root xylem through the length of the plant and to the leaves

trophic levels: steps in the food chain

tropism: involuntary response of an organism to an external stimulus such as light, water, gravity, or nutrients

vascular tissue: stem tissue of plants; includes xylem and phloem

vegetative propagation: an asexual process; asexual reproduction that occurs through mitosis only; does not involve gametes; produces offspring genetically identical to the parent

vertebrate: animal species with internal backbones

virus: structure containing a protein capsule, DNA, or RNA, and sometimes enzymes; can reproduce, but do not have the ability to conduct metabolic functions on their own; survive and replicate by invading a living cell then utilizing the cell's mechanisms to reproduce itself, sometimes destroying the cell in the process

xylem: vascular stem tissue of plants composed of long tubular cells, which transport water up from the ground to the branches and leaves

zygote: resultant diploid cell of sperm fertilizing an egg

Index

Biennial plants, 55
Bile, 81
Binary fission, 49
Binomial nomenclature, 135
Biogeochemical cycles, 100–105
Biogeography, 112
Biomass, 101
Biomedical progress, 117
Biomes, 110–113
Biosphere, 99
Biotic factors, 106
Biotic habitat, 100
Birds, 138
Birth rate, 105
Blastula, 84
Blood, 79
Blood tissue, 72
Blood type, 93–94
Bones, 76
Bone tissue, 72
Botany. *See* Plants/Plantae
Brain, 77, 78
Bronchi, 75
Bronchioles, 75
Bryophytes, 55
Buffers, 19
Bulbs, 66

C

C3 plants, 63
C4 plants, 63
Calvin Cycle, 41
Cambrian explosion, 131
Capillaries, 80
Carbohydrates, 20
Carbon cycle, 104
Carbon dioxide (CO_2)
  in carbon cycle, 104
  in photosynthesis, 40, 41
Cardiac muscle, 72, 76
Carnivores, 101
Carrying capacity, 108, 122
Cartilage tissue, 72
Catabolism, 40
Cell body, 77
Cell cycle, 47
Cell division, 45–51, 83–85
Cell membranes, 28, 29–32, 36

Cell structure and function, 27–29
Cell theory, 27
Cellular biology, 27–43
  animal cells, 32–36
  cell membrane properties, 29–32
  cell structure and function, 27–29
  plant cells, 36–43
Cellular metabolism, 40, 79–80
Cellular respiration, 41
Cellulose, 20
Cell walls, 36
Central nervous system (CNS), 77
Central vacuole, 37
Centrioles, 34
Centromere, 45
Cephalochordata, 138
Cerebellum, 79
Cerebrum, 79
Chaparral, 111
Charge, 13
Chemical bonds, 16–17
Chemical reactions, 17
Chemistry, defined, 13
Chemistry of biology, 13–23
Chlorophyll, 37, 41
Chloroplasts, 29, 37
Chondrichthyes, 138
Chordata, 138
Chorioallantoic membrane, 85
Chorion, 85
Chromatids, 45
Chromatin, 45
Chromosomes, 45–46, 90, 94
Circulatory system, 79–80
Class, 136
Classification of living organisms, 135–138
Cleavage, 84
CLEP Biology Exam
  about, 3
  accommodations, 7
  administration, 7
  format and content, 8–9
  military personnel and U.S. veterans, 7
  scoring, 8, 9
  study tips, 8–9, 10
  test-taking tips, 8–9
  when/where given, 3
Climax communities, 110

REA's Test Preps
The Best in Test Preparation

- REA "Test Preps" are **far more** comprehensive than any other test preparation series
- Each book contains full-length practice tests based on the most recent exams
- **Every** type of question likely to be given on the exams is included
- Answers are accompanied by **full** and **detailed** explanations

REA publishes hundreds of test prep books. Some of our titles include:

Advanced Placement Exams (APs)
Art History
Biology
Calculus AB & BC
Chemistry
Economics
English Language &
    Composition
English Literature &
    Composition
European History
French Language
Government & Politics
Latin Vergil
Physics B & C
Psychology
Spanish Language
Statistics
United States History
World History

**College-Level Examination
    Program (CLEP)**
American Government
College Algebra
General Examinations
History of the United States I
History of the United States II
Introduction to Educational
    Psychology
Human Growth and Development
Introductory Psychology
Introductory Sociology
Principles of Management
Principles of Marketing
Spanish
Western Civilization I
Western Civilization II

SAT Subject Tests
Biology E/M
Chemistry
French
German
Literature
Mathematics Level 1, 2
Physics
Spanish
United States History

Graduate Record Exams (GREs)
Biology
Chemistry
Computer Science
General
Literature in English
Mathematics
Physics
Psychology

ACT - ACT Assessment

ASVAB - Armed Services
    Vocational Aptitude Battery

CBEST - California Basic Educational Skills Test

CDL - Commercial Driver License
    Exam

COOP, HSPT & TACHS - Catholic High
    School Admission Tests

FE (EIT) - Fundamentals of
Engineering Exams

FTCE - Florida Teacher Certification
    Examinations

GED
GMAT - Graduate Management
    Admission Test
LSAT - Law School Admission Test
MAT - Miller Analogies Test
MCAT - Medical College Admission
    Test
MTEL - Massachusetts Tests for
Educator Licensure
NJ HSPA - New Jersey High School
    Proficiency Assessment
NYSTCE - New York State Teacher
    Certification Examinations
PRAXIS PLT - Principles of
    Learning & Teaching Tests
PRAXIS PPST - Pre-Professional
Skills Tests
PSAT/NMSQT
SAT
TExES - Texas Examinations of
    Educator Standards
THEA - Texas Higher Education
    Assessment
TOEFL - Test of English as a
Foreign Language
USMLE Steps 1,2,3 - U.S. Medical
    Licensing Exams

*For information about any of REA's
books, visit www.rea.com*

Research & Education Association
61 Ethel Road W., Piscataway, NJ 08854
Phone: (732) 819-8880